W9-AUA-239

MORE NUMBERS EVERY DAY

MORE NUMBERS EVERY DAY

HOW DATA, STATS, AND FIGURES CONTROL OUR LIVES AND HOW TO SET OURSELVES FREE

MICAEL DAHLEN AND
HELGE THORBJØRNSEN

TRANSLATED BY PAUL NORLEN

New York

Originally published in Sweden as *Sifferdjur* by Volante in 2021

Cover design by Sara Pinsonault

Hachette Books
Hachette Book Group
1290 Avenue of the Americas
New York, NY 10104
HachetteBooks.com
Twitter.com/HachetteBooks
Instagram.com/HachetteBooks

First Edition: March 2023

Published by Hachette Books, an imprint of Perseus Books, LLC, a subsidiary of Hachette Book Group, Inc. The Hachette Books name and logo is a trademark of the Hachette Book Group.

Library of Congress Cataloging-in-Publication Data has been applied for.

ISBNs: 978-0-306-83084-6 (hardcover); 978-0-306-83086-0 (ebook)

Library of Congress Control Number: 2022042495

Printed in the United States of America

LSC-C

Printing 1, 2022

CONTENTS

They look so innocent sitting there, those numbers. A lone number on a screen or a piece of paper. Your account balance, your pulse, or the number of steps before lunch.

1,590 97 3,467

Numbers are so concrete, exact, and clear. Numbers don't lie. They are honest, controllable, and neutral. A rational, enlightened society is based on numbers, not emotions. Numbers provide transparency, reliability, and evidence. Numbers are relevant, rational, and objective.

0 55 7.9

So we thought. But instead it turns out that they're flirtatious, manipulative, distracting little devils.

2 4 16

Numbers mislead us and lie. They distort and entice. They divide and command. They have slipped in everywhere you look—and are in the process of taking over your life. You love them; you depend on them. And they're in the process of really messing things up for you.

You just don't know it yet.

1 2 3

FOREWORD

Our days are numbered.

Literally. Everything we do during the day is counted. The days we socialize. The days we exercise. The days we work, study, travel. The nights we sleep. Our phones, social media, email, and apps count it all up, day by day.

How many steps have you taken today?

How many friends do you have?

How good is the driver in that car you ordered and are about to get into (what used to be called a "taxi" back in the day)?

You know all that because there are numbers for it. The pedometer counts the steps for you. Facebook counts your friends for you. The ride-share app calculates the average score of the driver for you.

Just a few years ago, you had no idea. But today there are counters for everything you do in a day. And at night too for that matter. If you want to know how long you've slept, how deeply, and how many times you woke up, snored, or turned over in bed (or were "social"), there are counters for that too. Search "counter" in the app store, and you'll get blisters from scrolling before you reach the end of the list. Do a Google search for counter apps, and you'll get over a million hits.

All those counters are a symptom that something is happening with our lives.

Not too long ago we managed to get through the day just fine without knowing how many steps we'd taken. We got along well with our friends without counting them. But just as soon as we had numbers for such things, those numbers suddenly became important to us. We started thinking about those numbers, feeling pleased by them, worrying about them, comparing them, and evaluating ourselves by them. They got us to take more steps. Make more friends. Get stressed about our low sleep numbers (and presumably lose even more sleep because of that). As if our lives depended on it.

This development is epidemic. We find ourselves in a place where numbers work their way more and more into our lives, into everything we do and are, influencing how we behave, what decisions we make, and how we think, sense, and feel.

Through the centuries we humans have been programmed to automatically, instinctively, react to numbers. Even if we truly wanted to stop using them, we probably wouldn't be able to. We are number animals. We have the same basic instincts as other animals, but what separates us from apes and cats, for example, is that our animal instincts are coded with numbers (even at the cellular level, as you will see).

However, human evolution probably didn't count on our having to deal with as many or as large numbers as we do now. According to estimates, we now generate collectively more numbers every day than all humankind combined scraped together between the creation, more than 5,000 years ago, of the first clay tablet in Uruk and 2010.

More. Numbers. Every. Day.

What is that actually doing to us?

We—Micael and Helge—started asking ourselves that question more and more often as we lectured and did research together about people's lives, behaviors, motivations, and happiness. We decided to find the answer. Or more precisely, the answers, in the plural. We have devoted several years to investigations, surveys, laboratory experiments, field studies, tests, interviews, and observations, and we have compiled the (often rather startling) results in this book.

Here you are going to find out how numbers affect you physically, to the extent that they get you to age slower or faster. How numbers affect your self-image and make you feel better or worse. How they color your experiences, even influencing how you feel pain. We're also going to show how numbers have come to determine how you perform and how they have even wheedled their way into your relationships.

Some effects are positive (e.g., numbers actually get you to perform better); quite a few are bad (e.g., you care less about what you're actually performing in). Some are slightly unpleasant (e.g., numbers can make you clinically depressed), and many of them are funny (e.g., certain numbers make you more inclined to turn left). Our hope is that this book will help you to become aware of all these effects so that you can take the good into account, counteract the bad, and hopefully never need to experience the unpleasant. So that you feel more content, have more fulfilling experiences, get more out of your relationships (your partner, current or future, is going to thank you), and live a healthier life.

You are also going to get some good stories to tell. About how a certain jersey number was required for Michael Jordan to become the GOAT (that is, the "greatest of all time," as basketball aficionados say). How step counters can create a housing bubble. Why the probability of getting a parking ticket is much higher right before Christmas than during the rest of the year. How a book about the genetics of flies became the world's most expensive within the course of 24 hours. Or what Jesus and Kim Jong Il have in common and how this has affected millions of people's lives.

"But wait, there's more," as they used to say on the TV shopping channels. We are also going to look more closely at how the number epidemic affects us not just as individuals (which itself has a big impact) but as an entire society. Numbers have also worked their way more and more into politics. As soon as it became possible for politicians to get numbers on how many people saw and internalized their message, they started adapting the message in real time to get those numbers as high as they could: maximizing their crowd appeal, making more promises, being more provocative, and acting more and more like caricatures. Build walls (or in any event promise/threaten to do so) instead of bridges. You probably know whom we're referring to: Donald Trump was a clear symptom of the number epidemic. The campaign that made him president was led by numbers as algorithms guided the message based on what produced the most clicks and shares.

Numbers become a truth that influences decisions on all possible levels in society, both in companies and in the public sector. Things that are easier to measure get prioritized, like the strength of the lighting in a workplace instead of the employees' well-being, to take a funny example we'll come back to.

For us, as professors of economics, it also seems obvious to point out that increasing access to numbers is in the process of making them a currency in themselves. A currency we can exchange with one another and with which we can trade and bargain. Likes, swipes, scores, and points are all types of useful behavioral data. In one regard we can see this as a positive—a possible alternative to money with the potential to even out the differences between poor and rich and give everyone the chance to create their own capital—making it profitable, for instance, to be kind, have friends, and share. But what happens when the numbers simply become a new sort of money, with all its flaws too? When it suddenly becomes possible to put a price on friendship? Buy and sell likes? The risk is that we will become number capitalists, greedy for more and bigger numbers. Or even that we will act in immoral ways. Amusingly, one of our studies shows that people who get an unusually large number of likes on an Instagram picture also become more inclined to steal printer paper at work.

We are going to show how numbers can make us depressed, narcissistic, and immoral but also motivated, strong, and enthusiastic. You will learn how certain numbers stick in your brain and unconsciously influence what you are willing to pay for a house, a car, or a bottle of wine. And we're going to explain how we humans sometimes relate to numbers as if they had a personality and gender all their own.

Numbers can be dangerous but also amazing, and our goal is not to get you to stop using them. We, of course, love numbers (otherwise we never would have survived so long as economics professors). They're one of humanity's most important inventions, the first thing people considered worth writing down at all, according to archaeologists. The world's first-known text is a clay tablet, excavated in what was once Mesopotamia and dated to 3200 BC, accounting for the number of goods and assets of different types in the temple of the capital Uruk. A sort of Excel spreadsheet in clay, in other words.

Since then numbers have followed us through history. Not only in accounting but also in culture, religion, language, time, and civilization. But in recent years our use of numbers has exploded.

Have we been *numberfied?*

Exponential technological development has made it possible for us to generate numbers to a much larger extent than just a few years ago. The calculating power in our computers has increased 60 to 100 times time since just 2010, according to the TOP500 list. That means that, following Moore's law, it has increased by a magnitude of 20 to 200 percent every single year for the past 50 years (depending on how you count it). Just three decades ago capacity corresponding to the computers we have in our hands, which are called "phones" nowadays, would have cost over $100,000. We can fill these phones with calculators and apps, and there are computer systems, servers, and clouds that can register and store everything we do around the clock everywhere, all the time.

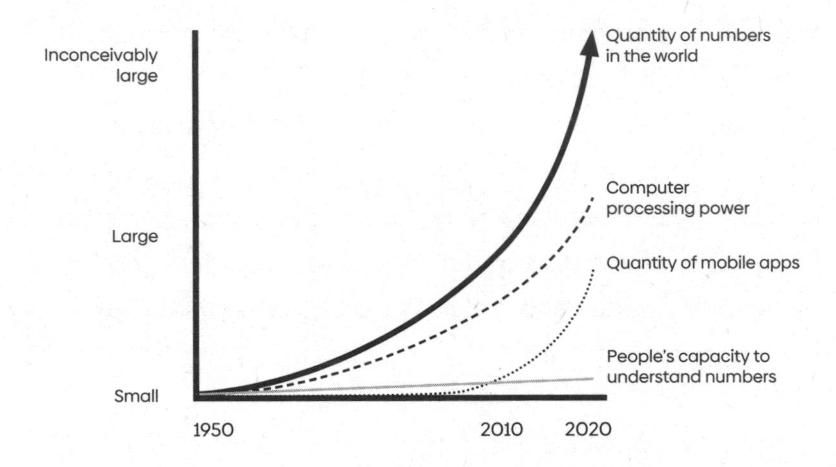

A few years ago, we—Micael and Helge—stood together on a stage and asked 500 company executives a simple question: "What would be harder for you—a week without alcohol, without sex, without friends, without money, or without your phone?" And the result was both crystal clear and sad: a week without their phone was the most painful thing these executives could imagine.

But that's not really so strange, is it? Little by little we've let our phones into every intimate part of our private lives. Our health, our money, our job, our friends, our vacations. All of it. And in exchange the technology gives us continuous injections of what we all have become completely dependent on: numbers. Numbers for everything and everyone, in all forms and variations.

An additional explanation for the number epidemic is that

we have a surplus of things in our lives now to put numbers on. We have more and do more. According to American statistics, people's average living space has almost tripled in recent decades, consumption has more than doubled, and Americans spend more than $24 billion annually just on storage space for all the stuff they have. We have more careers and change jobs more often (according to the US Bureau of Labor Statistics, people stay on average four years in the same job, far from the gold-watch standard of yesteryear). Simultaneously, free time has increased by approximately two hours a week in the OECD (Organization for Economic Cooperation and Development) countries and by almost twice as much in our native Norway and Sweden. And that was before the coronavirus and the subsequent increase in remote working. As if that weren't enough, nowadays we have more waking hours in the day to fill with numbers: according to Finnish researchers, they have increased from 16 to 17 on average the past decade while an American study found that the percentage of people who sleep at most 6 hours (and thus have 18 waking hours) has increased by 30 percent in recent decades.

With this surplus also comes increased uncertainty. In *Nextopia* (2008), Micael coined the term "the -ever world," where *whatever* has become available for *whoever, wherever,* and *whenever.* Back then, a search on Google to "buy shoes" produced 500,000 hits. Today the same search produces almost six million hits. Whatever we're looking for, whether it's things to buy, trainings to choose, jobs to apply for, recreational activities to spend time on, restaurants to try, drivers to ride with, or people to date, there are 10 times as many offerings. How are we ever able to choose?

This almost certainly contributes to the reduced number of sleep hours in the statistic because of sleep disorders and increased stress, not least among young people. And clearly it also feeds back into our increasing reliance on numbers as more ratings and better scores help relieve our decision-making anxiety.

The growing surplus of things to quantify also means increased competition for our attention. Numbers become a decisive tool that provides authority. Companies fill their marketing with them to get us to stop and choose their products (a paddle tennis racket that gives 27 degrees more spin sounds good, whatever that actually means). The news media spice their headlines with numbers to get us to read their articles ("Covid-19 deaths increase by 100 percent in a week"). Politicians use them to sell their policies ("We have boosted the economy by building 30,000 new homes!"). We ourselves use them in everything from selling our used clothing to renting out our couch to dating, in the hope that people will choose just us because of our high averages. The numbers don't demand long explanations and aren't subjective (we think). We react to them instinctively and understand them immediately (we think).

And so this is where we find ourselves now.

Our days are numbered, literally speaking.

This doesn't have to mean that our days are also numbered, figuratively speaking. There are worse threats to human existence than the number epidemic. (Viral pandemics, for example. The climate threat. The hundreds of thousands of asteroids that pass through our solar system and risk colliding with the Earth—actually forget it, those examples probably don't make you feel much better...) But in counting our days, do we perhaps make our existence a little poorer?

With this book we don't want to save the world from numbers, but we do want to bring your attention to how you are influenced by them and to help you relate to them so that the quantification doesn't make your life poorer. Perhaps you'll decide that some parts of your life can actually be dequantified. Or that, at the very least, they could do with a temporary number detox. In any event, we think that everyone would feel better for getting vaccinated against numbers, so as to be able to choose for themselves how to handle them.

Look at this book as your number vaccine.

MORE NUMBERS EVERY DAY

THE HISTORY OF NUMBERS

First let's rewind the tape a little.

The first "spreadsheet" with accounting from the temple in Uruk may be dated to 3200 BC, but numbers are considerably older than that. Their history actually started over 40,000 years ago. Imagine that. Because that's how old the oldest counting sticks that archaeologists have found are. Made from bone, they are the first sure sign that humans had started to count—initiating all the remarkable things that would come in the wake of this.

I I I I I I I I I I I I I I I I I I I I I I I I I I I I I

The so-called Lebombo bone, which was found in the 1970s in the mountains in Swaziland, had 29 cuts. Some argue this may indicate that African women were the first mathematicians and that they used counting sticks to keep track of their menstrual cycles. Whether that's true we will never know, because the bone was broken off after the twenty-ninth line. Perhaps it had been even longer?

Really old counting sticks have been found in Europe as well. The famous "wolf bone," found in Czechoslovakia in 1937, is estimated to be about 30,000 years old. The bone has a total of 55 counting marks arranged in groups of 5 marks each.

IIIII IIIII IIIII IIIII IIIII IIIII IIIII IIIII IIIII IIIII IIIII

The wolf bone can in many ways be considered humanity's very first supercomputer. With such a counting stick, we could both count and write down numbers, enabling us to get an overview and create order. We could keep track of the number of individuals in a flock, tally our animals and possessions, and later even do calculations in connection with trade. People across the world developed, slowly but surely, the capacity to count and make calculations and started to ascribe meaning and value to numbers.

We soon became dependent on numbers, in part because they came to be quite necessary for governing societies and conducting trade. The first writing tablet from Mesopotamia illustrates just this: It consists of jottings of numbers and calculations. And hey, presto! Economists were born. *Four for you and five for me.*

Now humans didn't invent numbers. They were already there, of course. For anyone who wants to count, nature, including the human body, is a gold mine. Fingers and toes, animals, eggs. Those were presumably the first things people started counting. Other numbers and patterns in nature are a bit more complicated and harder to see, such as pi, or the Fibonacci sequence, which is actually a spiral. But if you carefully observe seeds in a pinecone, you will discover that they too are arranged in spirals: 5 spirals in one direction and 8 spirals in the other. Sunflowers also have their seeds arranged in spirals: 21 in one direction, 34 in the other. Count for yourself, and you'll see. And if you look carefully at the Romanesco variety of broccoli the next time you're in the grocery store, you can also find and count the Fibonacci spirals. In mathematical terms it's an amazing vegetable, which, just like the rest of nature, is filled to the brim with numbers and patterns.

1, 1, 2, 3, 5, 8, 13, 21, 34, 55, 89, 144, 233, 377...

Speaking of Fibonacci, when you learn about that concept in school, it can really make your head spin. I remember that as an enthusiastic high school student in the 1980s,

I suddenly started searching for spirals and number sequences everywhere. And those who seek shall find. Petals on flowers? Fibonacci. Patterns on grungy T-shirts? Fibonacci. Pineapple (which was very popular in the 1980s, even on pizza)? Fibonacci. The shape of the ear, galaxies—everything. Fibonacci.

Even the golden mean, which we learned about in art class, proved to be about Fibonacci. We learned that humans perceive the golden mean as something fine and harmonious. By using mini-calculators and rulers, we figured out how artists throughout the ages have used the golden mean in their compositions to create something beautiful.

Maybe our teacher also got tunnel vision (or spiral vision?) in the end where Fibonacci was concerned. Because we had the same teacher for PE, we got a bit of a hybrid exercise: measuring the golden mean on the length of ourselves. And if you're wondering: for the majority in the class, the golden mean was exactly in the middle of the navel. Except on poor Kristian with his long legs.

Helge

Anthropologists think that the origin of our understanding of numbers is our fascination with our own hands—5 fingers on each hand. In many societies the discovery that "a hand is 5 things" was a catalyst for much faster development.

Think about it! Someone just looked at their hand, thought a little, discussed with a friend—and just like that you got a kickstart for understanding everything from numbers and trade to maps. It's extremely intuitive and simple to count using your fingers. Both children and adults do it even today. The number of fingers and toes was also the origin of many early cultures' number systems, based on the numbers 5 and 10. Just like with the wolf bone.

With the discovery of numbers, we humans could suddenly show one another quantity and trade with each other, calculate profit, do accounting, and even introduce taxes and fees. We pulled away from other species at record speed. Zoologists think that certain other mammals actually have the capacity to count up to 3 or 4, but that's peanuts compared with our ancestors, who suddenly managed both 5 and 5,000.

Numbers and understanding of numbers became incredibly important when we humans gradually started engaging in trade and when we organized our societies and started to live more densely. The ability to count is also a prerequisite for greed, negotiations, and status. If you're going to get anywhere in life, you must be able to count and compare. For that reason, societies through the ages have had different numerical systems, each of which has developed its own sort of rhythm or base. Our decimal system—or the Hindu-Arabic numeral system, as it's also called—has the basic rhythm of 10. The binary number system, which all modern computers use, has 2 as its basic rhythm. Everything is written in combinations of two numbers: 0 and 1. Ancient Babylon, amusingly enough, had a number system with 60 as the basic rhythm. That system became important for calculating time—seconds, minutes, and hours—and also for measuring angles in a circle. In just about any other context, the Babylonian number system is pretty impractical. It didn't even have a sign for zero.

Throughout history we have had a series of different number systems with basic rhythms of 5 and 10—yes, based on the number of fingers and toes, which we humans gradually discovered that we had. Intuitively it's easy to understand how these number systems arose, isn't it? Roman numerals are based on the basic rhythm 5, where V is 5 and L is 50. But that number system is also extremely complicated and difficult to navigate. Just take a look at old clocks and calendars. It is, after all, the MMXXs now.

Incidentally, the Romans floundered in the development of both numbers and mathematics in the world. When they invaded Greece, they were interested in power, not numbers. The Roman numeral system was far too complicated for use in counting and calculating, but it worked nicely to keep track of how many men had been killed. When the Romans killed the Greek mathematician and inventor Archimedes and introduced the Roman numeral system, the development of both mathematics and other sciences slowed down considerably. Roman numerals spread over all of Europe and actually became the dominant number system for more than 500 years. And yet, do you remember the names of any Roman mathematicians? You don't? That's not so strange. They simply weren't very good.

As an economics professor I often think about numbers as language, to be able to communicate, plan, and agree on how we should use, share, and trade various resources. In light of that, it's actually quite fascinating that humanity (or a large proportion of it) is in agreement on the same number language. I mean, how many languages are there in the world? I checked on Wikipedia and saw that there are more than 100 languages that are spoken by at least five million people. That must say something about how intuitively we use numbers?

Personally, however, I'm not convinced that the number system we use today is the best possible option. I'm quite charmed by the number system that the Cistercian monks used in medieval times in France, which had different counting lines for ones, tens, hundreds, and so on. Everyone who has tried an advanced mental calculation knows that it's a much faster and more effective system.

Micael

Fortunately the Roman Empire fell at last, and people could return to the more sensible Hindu-Arabic decimal number system. Then people's innovation capacity (and need for counting) could again blossom and grow.

And it grew. The innovation capacity, that is. Substantially.

Numbers and mathematics enabled humans to achieve astounding things. Numbers are behind everything from the pyramids and the first trip to the moon to every new computer or smartphone in the world. Here is where we come to what is making the number epidemic so dangerous and important right now—the deadly cocktail, if you will: people's innate fascination with and dependence on numbers combined with the fact that technology has made numbers ubiquitous. They're everywhere! And whether you hate or love mathematics, those numbers have power over you. Common to all numbers and number systems is, namely, that they have, and have always had, a massive influence on people's thoughts, beliefs, and superstitions.

> Sorry, but I'm having a hard time letting go of the thought that the number system we use today is perhaps not the best one. I was at a conference a few years ago where two British professors of computer science launched a new system that they called "interactive numbers." It's not that easy to explain—I still don't understand it completely myself—but roughly speaking, it's based on the fact that

digital numbers (and basically all numbers are of course digital nowadays) should be able to correct themselves while we enter them based on how reasonable they are in relation to other numbers we have entered. Because the problem is that we often make mistakes, compared with when we used to write them by hand: we press the wrong number, happen to hold the key too long so that the number is doubled, miss a space, enter a comma wrong, or whatever. An eye movement measurement showed that people who enter numbers devote 91 percent of their attention to the keypad and only 9 percent to the numbers on the screen.

One of the examples they used was from Norway: In 2007 Grete Fossbakken lost the 500,000 kroner that she meant to transfer to her daughter's bank account; the money ended up somewhere else altogether when she made a keyboard error. Evidently this happens in 0.2 percent of all bank transactions (and added up, that's a rather large amount of money . . .). Another example is the British citizen Nigel Lang who in 2011 was arrested on suspicion of sharing indecent images of children, but no such images were found on his computer. Much later it was discovered that the police had mistakenly added a digit to the IP address on the computer they were searching for. He received £60,000 in damages plus court costs.

Micael

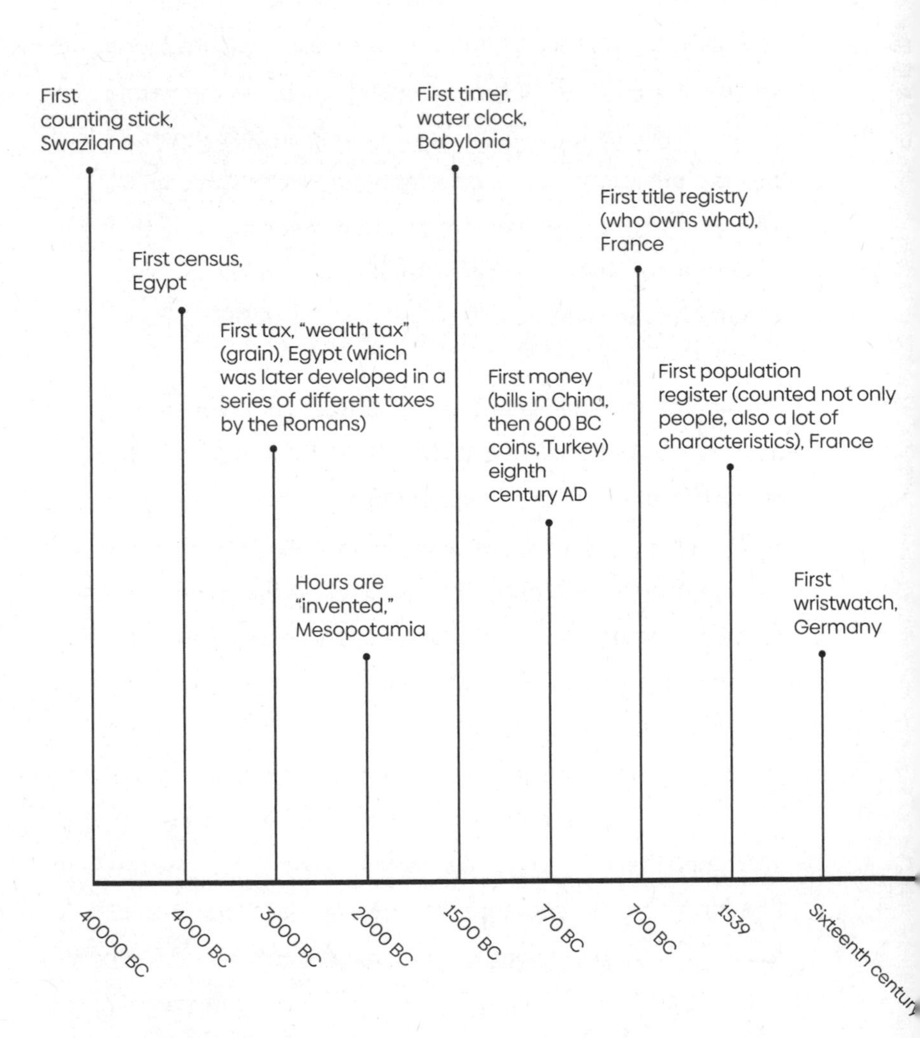
First
counting stick,
Swaziland
First census,
Egypt
First tax, "wealth tax"
(grain), Egypt (which
was later developed in a
series of different taxes
by the Romans)
Hours are
"invented,"
Mesopotamia
First timer,
water clock,
Babylonia
First money
(bills in China,
then 600 BC
coins, Turkey)
eighth
century AD
First title registry
(who owns what),
France
First population
register (counted not only
people, also a lot of
characteristics), France
First
wristwatch,
Germany
40000 BC
4000 BC
3000 BC
2000 BC
1500 BC
770 BC
700 BC
1539
Sixteenth century

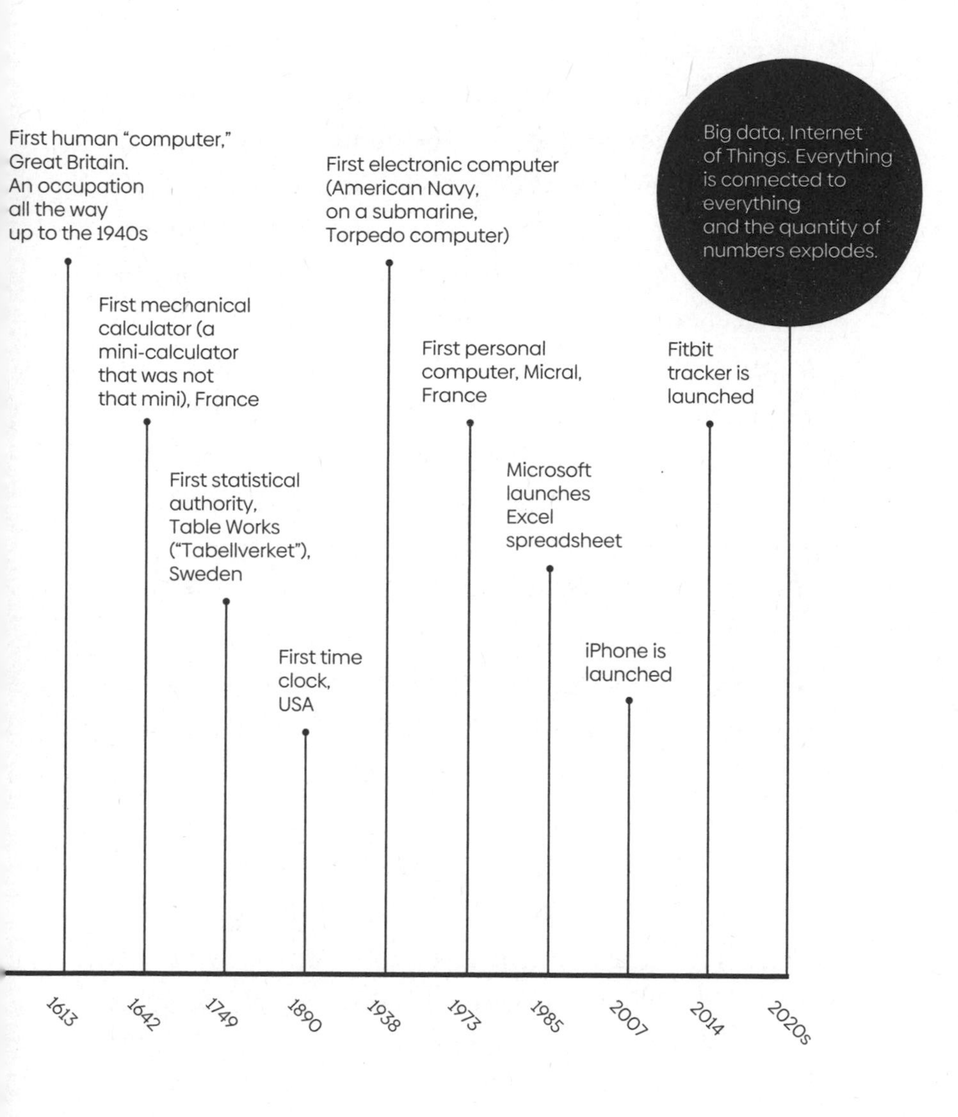
First human "computer," Great Britain. An occupation all the way up to the 1940s
First mechanical calculator (a mini-calculator that was not that mini), France
First statistical authority, Table Works ("Tabellverket"), Sweden
First time clock, USA
First electronic computer (American Navy, on a submarine, Torpedo computer)
First personal computer, Micral, France
Microsoft launches Excel spreadsheet
iPhone is launched
Fitbit tracker is launched
Big data, Internet of Things. Everything is connected to everything and the quantity of numbers explodes.
1613
1642
1749
1890
1938
1973
1985
2007
2014
2020s

THE MYSTIQUE OF NUMBERS

We humans see figures and numbers everywhere: in words, signs, names, clouds, and nature. We find context where we *want* to see context and ascribe the numbers important significance whether they occur in the news or in social media, in nature, or on lottery tickets. Certain digits and numbers also become important to us and gain a significance and symbolism of their own.

A prime example is the number 666 from the book of Revelation in the Bible, also called the "number of the beast." Throughout history countless people have linked this number to odious persons in their own times and assigned them the role of the Antichrist incarnate.

We also ascribe certain significance to other numbers: 13 means bad luck, 3 is sacred, and 1,000 is a lot. Some numbers are so closely connected to certain events or concepts that they become almost magical. The belief in a sacred or mystical connection between numbers and events even has its own name: numerology. Have you read *The Da Vinci Code* by Dan Brown or seen the movie? A professor of semiotics and a cryptologist solve a mathematical puzzle linked to the murder of a curator at the Louvre. The movie—which, by the way, the Catholic Church criticized harshly—depicts countless instances of numerology. Whether it concerns the Fibonacci sequence, the Hebrew system, or other number systems, numerology has played a role in just about every culture.

History is full of numbers and number magic. Alchemists, philosophers, religious leaders, and even doctors have been inspired by the mystical aura around numbers. For example, doctors of traditional Chinese medicine and similar practices such as acupuncture base their systems on mystical numerical connections, like "365 parts of the body, one for every day of the year" and "twelve paths where blood and air circulate, just as the twelve rivers flow into the central kingdom." And even if the church has, on occasion, strongly opposed numerology, that sort of thing is found in both the Bible and other religious texts. For example, the numbers 3 and 7 have a strong spiritual presence in the Bible. God created the world in 7 days. Jesus asked God 3 times if he could avoid the crucifixion, and he was crucified at 3 o'clock in the afternoon.

In applications of Islam and Islamic astrology, the number 7 similarly plays an important role. Seven was originally the number of planets and is the first "whole" number because it can consist of 3 + 4 as well as 2 + 5 and 1 + 6; it is therefore also the sum of the dots on the opposite sides of a die. In the Koran, there are 7 heavens, the first chapter has 7 verses, and the pilgrims in Mecca walked 7 times around the Kaaba and cast 7 stones at the walls that represented the devil.

In Judaism and Buddhism we also find numerology closely connected to religion in ancient times. In Jewish mysticism, especially in the Kabbalah, numerology and astrology are very prominent. Tenacious Kabbalists thought that the Old

Testament was written in a code inspired by God. Their system of numerology was an attempt to decode the scriptures.

The Kabbalah has inspired not only Christian mystics but also commercial New Age movements such as Philip Berg's "Kabbalah" cult, which incidentally has attracted the likes of Madonna, Guy Ritchie, and Demi Moore.

During the medieval period, the "science" of arithmology—a kind of philosophy linked to numerology and to the power and symbolism of numbers—was developed and often used by the Christian leaders and artists of the time. For example, the Italian poet Dante Alighieri's works are filled with number patterns and number symbolism. His famous *The Divine Comedy* is based to a high degree on the number 3 and the Trinity. The number 3 shows up throughout the work: 3 parts, 33 songs, 33 verses, 3 verse lines. The devil has 3 faces; 3 women pray for Dante; there are 3 frightful monsters and 3 kingdoms after death. Through the entire medieval period and the Renaissance, number mysticism plays a prominent role, with numerous books making use of numbers and numerical systems, as Dante did, or developing arithmology and numerology as a kind of superknowledge that would unite all other sciences.

THE FATHER OF THE NUMBER EPIDEMIC

People's historical fascination with both numbers and numerology thus seems to concern a cockeyed mixture of mathematics, philosophy, religion, art, astrology, and mysticism. Interestingly, very many of these thoughts and movements can be traced back to a single operator: a man by the name of Pythagoras.

Do you remember him from your math classes in school? Most people who've taken geometry will recall the Pythagorean theorem about the length of the sides of a right triangle. Not as many know that he was a mathematician, philosopher, and mystic who lived 500 years before Christ—and founded a whole movement and a school of mystery. His thinking has greatly influenced Western philosophy, mathematics, music, and religion, and his ideas inspired philosophers such as Plato and Socrates, as well as astrologers, musicians, and adherents of Kabbalah. Pythagoras maintained that all things are basically mathematical and can be understood as numbers, and he lectured on mathematical connections in everything from music, geometry, and astrology to nature, such as the seven colors of the rainbow and the Earth's five climate zones. He preached the beauty and logic in harmonies consisting of whole numbers.

Pythagoras was a legend already during his lifetime and, according to Aristotle, was an almost supernatural person. For that reason he quickly attracted a large following, whose members later came to be called Pythagoreans. These adherents, who were ascetic and temperate and devoted themselves to mathematics, music, and astronomy, came up with many mysterious things. His disciple Hippasus was drowned, for example, because he thought that the quadratic root of 2 was not a rational number, and his followers also became extremely interested in the differences between odd and even numbers.

And maybe they were onto something. More recent research in numerical cognition, which we'll come back to shortly, has shown that even numbers are perceived as feminine and soft, while odd numbers are perceived as masculine and hard. Over 2,000 years ago, the Pythagoreans sat in white, ankle-length robes and argued the exact same thing: odd numbers were masculine, even numbers feminine.

The Pythagoreans, all of whom were certainly men, thought, however, that the masculine odd numbers could be linked to what is light and good, while the feminine even numbers could be linked to things dark and evil. For just that reason, even numbers were not especially popular for several centuries. For Plato even numbers were a bad omen. The Talmud contains lots of examples of the use of odd numbers and avoidance of even numbers. Mohammed also clearly preferred odd numbers, and ancient physicians and doctors always gave their patients an odd number of tablets. And which numbers turned out to be the most important in most religions? Yes, the odd numbers 3 and 7.

Does this mean that even today we like such numbers better?

THE NUMBERS WE LOVE AND HATE

Are you the sort of person who feels a little uneasy when the volume on the remote control shows 43 instead of 44 or 42? Or do you think that the number 20 feels calmer and softer than 19? Then you're hardly alone. Odd numbers are a little more individualistic, restless, and difficult, many people think. Even numbers are friendly, uncontroversial, and easier to understand. The number 10 is good, 11 is tricky. Research has shown that odd numbers are challenging because the brain needs a little more time to process them. Even numbers flow easily into the brain and are easily processed. Odd numbers trip the brain up more.

These days we know quite a bit about which numbers the brain likes and which are trickier and more difficult. There are also both simple and quite intricate explanations for why we humans experience different numbers in different ways. An extensive study from 2020 gives a slightly special explanation for why we perceive divisible numbers (such as 4) and indivisible numbers, also called prime numbers (such as 5), so differently. You see, we ascribe numbers human characteristics and relate to them accordingly—a little like we do with objects and brand-name goods. Certain objects are masculine; others are feminine. Certain brands are sophisticated; others are cruder. Our thinking is the same where numbers are concerned: divisible numbers have connections to many other numbers and are perceived as social, while indivisible (prime) numbers lack connections to other numbers and are perceived as solitary.

Researchers also show how this leads to our judging products and brands differently based on how we perceive the numbers linked to them. If you call a new car Audi A7, it will be perceived as solitary and individualistic. If you call the same car Audi A6, it will be perceived as more social. And vice versa: if you, as a consumer, are alone when making a decision, the probability is greater that you will choose a product, a characteristic, or a price with divisible numbers because you then have a greater inclination to choose something social. Single people actually prefer social, even numbers. Strange, isn't it? But scientifically pretty well documented.

And as said, research also shows exactly what the Pythagoreans maintained: that numbers also have gender. In a famous study from 2011, two researchers from Northwestern University in Evanston, Illinois, found that even numbers are perceived as feminine and soft to a higher degree, while odd numbers are perceived as masculine, independent, and strong. The researchers showed the participants in the study various foreign names that they couldn't identify in advance as belonging to a girl or a boy, and then they connected either an even or an odd number to the names. It turned out that the participants thought that the names were feminine more often when paired with an even number and masculine more often when paired with an odd number.

In a follow-up study the participants were shown random

images of babies, and then a number was linked to each. Once again the same pattern arose: baby pictures linked to an even number were more often assumed to be of girls, and baby pictures connected to odd numbers were more frequently assumed to be of boys. The subjects were actually 10 percent more inclined to think that the same baby was a boy than a girl if the picture was next to an odd number.

It also turns out that we humans have favorites among all these feminine or masculine, solitary or social numbers. A few years ago, Alex Bellos, author of *Alex's Adventures in Numberland* and *The Guardian*'s math blog, did an investigation on the internet to identify people's favorite numbers. The investigation in his selection showed that odd numbers overall are a bit more popular than even numbers. So despite the fact that we think odd numbers feel more uncomfortable and difficult than even ones, we like them better. Why? Perhaps just because the major world religions, inspired by the Pythagoreans, have always favored the masculine, odd numbers over the female, even ones. A kind of number chauvinism, if you will.

So which number was crowned the world's favorite? A total of 44,000 people sent in their favorite numbers, and just over half of these were numbers between 1 and 10. And the winner was—ta-da!—the number 7. Considering 7's presence in almost all religions and cultures, this is no great surprise. The number 7 pops up just about everywhere: 7 days, 7 deadly sins, 7 mountains, 7 brides, 7 fairy tales, 7 sisters, 7 seas, and 7 miracles. And the 7 dwarves, of course.

And yes, in second place came the number 3, which of course is also strongly integrated into most religions, signals both trinity and completion, and is considered a sacred number. The number 8 took third place, probably mostly because it signifies luck in China. The lucky number 8 is important to many Chinese people, and for just that reason the opening ceremony at the 2008 Summer Olympic Games in Beijing began 8 seconds and 8 minutes past 8 o'clock on the eighth of August (in Chinese literally called "number 8 month").

Now the number 0 was not included in the selection, but if it had been, the competition would have been closer. Ever since the Indian mathematician Brahmagupta formally introduced the number 0 in his work *Brāhmasphuṭasiddhānta* (try committing that to memory!) in AD 628, we have had a marvelous concept to understand absolutely nothing. Zero is just nothing, not an iota. We've become so enamored of zero that we've even given it nicknames such as "zip," "zilch," "nada," and "scratch." Even in sports we've created words such as "duck" (cricket), "nil" (soccer), and "love" (tennis) to mean zero.

NUMEROLOGY AND IDIOCY

The belief in a holy or mystical and meaningful connection between numbers and events has existed since even before Pythagoras but might seem incomprehensible for the majority of independent thinkers in modern times. Numerological fanatics are nonetheless found all over the world even today, and a lot of fascinating self-help books are sold on numerology. The logic in many of them is that all people have (or are) a separate number, which is calculated in a certain way. That number affects everything in your life, and you should also let the number decide where you live, what lottery numbers you play, where you travel, what hotel room you stay in, and what you name your child and your cat.

Some of these numerology books are truly great entertainment. Here is a short extract from the best-selling *Glynis Has Your Number* by Glynis McCants, a rich and famous numerologist who has been a guest on TV programs such as *60 Minutes*, *The Ricki Lake Show*, and *Dr. Phil*.

> Both my Birth Day Number and my Life Path Number add up to a 3 vibration, which makes me a double 3. On [one] trip I flew out on flight 33. Interesting, I thought. Then they put me in the 12th row—which, you'll soon learn, breaks down to a 3 as well. When I arrived at my hotel, I was on the 21st floor. Can you guess? Another 3. Then when I flew back, they put me in seat 30—a 3 again—and I wondered what the heck was going on. When the pilot announced that we would be flying at 33,000 feet, I just laughed! It's so fascinating how often the energy speaks to you.

Now it's not particularly hard to understand that anyone who constantly searches for threes in life will find them everywhere. A tricky thing with numbers and number patterns is discovering which are random and which are systematic or done with intent. Do people, especially "numerologists," perhaps have an exaggerated tendency to see connections where there aren't any? Consider Dante's *The Divine Comedy*, which we took earlier as an example of a book brimming with number patterns. Besides the obvious patterns in the verse form and the songs, both numerologists and academics have found a number of other prominent number patterns and number connections in the text. Did Dante do all this intentionally, or could some of it simply be down to chance? In an amusing and elegant article titled "Numerology and Probability in Dante," mathematics professor Richard Pegis does an analysis of these connections and finds—not surprisingly, perhaps—that the probability that they are random is approximately as great as if Dante had tossed a coin.

It's relatively easy of course to ridicule both medieval and contemporary commercial self-help numerology. We know better after all, modern, enlightened, and intelligent as we are. But doesn't a little congenital numerologist dwell inside all of us anyway? Maybe we have an insidious numerological lopsidedness in the brain, which means that we fear and avoid the number 13, bet on the same lottery sequence week after week, like 3 and 7 more than 4, and let ourselves be amused and guided by numbers every day.

And because there are now numbers everywhere, we are influenced by them both more often and more powerfully than we suspect. Never before has humankind produced more numbers. Exponentially and digitally. Yes, epidemically. The numbers are in every nook and cranny of your life and deep inside your brain. Numbers follow you to work, on vacation, to the toilet, and into bed.

Maybe the numbers have even sneaked all the way into your body?

NUMBERS AND THE BODY

"Number 45 is not number 23." That's how Nick Anderson of the Orlando Magic explained why he could steal the ball from Michael Jordan six seconds before the end of the first semifinal game against the Chicago Bulls. It was a historic moment. The year was 1995, and Michael Jordan was back with the team, which had won three straight National Basketball Association (NBA) championships before his one-year break. The world's best player returned to the world's best team, and it was time to take back the title the team had lost while he was away. But instead Nick Anderson stole the ball from Michael as he was about to take the last shot of the game. Nick passed it to teammate Horace Grant, who dunked in the two final game-winning points for Orlando. "I never would have been able to do that against number 23," said Nick Anderson, referring to the jersey number Michael Jordan had worn when the Bulls won the previous three championships. But in his comeback Michael chose jersey number 45 instead; suddenly he was no longer the world's best player on the world's best team, and the Chicago Bulls were eliminated in the semifinals.

The following season Michael changed back to number 23 and once again became the world's best player. And the Chicago Bulls won both the semifinals and the finals—three years in a row, again.

To say that the jersey number made Michael Jordan the world's best basketball player is perhaps giving it a bit too much significance. On the other hand, at this point you know that we humans have an inclination to read great significance into numbers and are influenced by them in every conceivable context. And sports are filled with numbers, not least in the United States, where fans, broadcast media, and betting companies collect statistics on everything.

For example, there are statistics showing that Michael Jordan averaged 27.5 points per game wearing jersey number 45—which isn't bad, but significantly less than the 31.0 points he made with number 23 on his jersey. There are also statistics showing that players with lower jersey numbers average more points per game than those with higher jersey numbers. This is in contrast to hockey, where the relationship is the opposite (the all-time scoring leader, Wayne Gretzky, wears 99, while many goalies, who almost never score, wear 1). According to statistics, it's better to have a jersey number under 50 in the NBA (preferably 31, which is the number with the highest average number of points) and over 50 in the National Hockey League (91 has the highest point average). Both leagues have in common that almost all players prefer to have jersey numbers with odd numbers instead of even.

We've circled back again to this business with odd and even numbers, where odd numbers are perceived as more masculine and even numbers as more feminine. In light of this, it's perhaps expected that so many male athletes choose odd numbers. The exception of course is the extremely sought-after number 10 in soccer, which has had a special symbolic meaning ever since the legendary player Pelé wore that jersey for the first time when Brazil won the World Cup against Sweden in 1958. Incidentally, he got that number by mistake; at that time jersey numbers were linked to the player's position on the pitch and number 10 was a midfielder's number, whereas Pelé was an attacker. But after the World Cup victory, Pelé refused to change his jersey number, and the rest is history. Regardless, the question remains this: Is it actually possible that numbers have an effect on us on a physical—as well as a psychological—level?

This chapter looks more closely at how our human bodies are invaded by numbers, influencing how strong we are, how we age, and how we move. In fact, we have reprogrammed a primitive part of the brain that we share with other animals on this planet to automatically react to numbers. Yes, we have become number animals.

MAGIC NUMBER BOUNDARIES

In one study, American college football players had to perform a classic strength test used by professional players in the National Football League—namely, to bench-press 225 pounds. They did this three times over a three-week period. Not surprisingly, on average they managed the same number of lifts on each occasion (in other words, none of the football players benefited from a wildly miraculous increase in strength). But what the players didn't know was that on one of those three occasions, the weight was only 215 pounds—the test leaders had deliberately mislabeled it. So, half the players lifted the correct weight the first week and the lighter weight the following week, and the other half lifted the lighter weight first and then the correct, heavier weight second. For both groups, the number of lifts they did went unchanged, meaning that the actual weight they were bench-pressing had no discernable effect at all. It didn't matter if it was 225 or 215 pounds—they were just as strong either way. Apparently, numbers are literally heavier than iron—at least 10 pounds heavier.

That numbers can weigh more than iron is also an

explanation for why it's so much harder to increase the weight you're bench-pressing from, say, 97.5 to 100 kilograms than to raise it from 95 to 97.5 kilograms. The difference in weight is the same, 2.5 kilograms, but the difference in numbers is greater when the 9 is replaced by a 10. Surely you've felt this at some point at the gym? The numbers that it can feel almost impossible to surpass are called sticking points, or "magic boundaries," but as soon as you have managed to do so, it gets considerably easier to raise them again. It's the number that influences your progress. In Norway and Sweden weightlifters get stuck at 100 kilograms, but an American might get stuck at 102.27 kilograms (or 225 pounds) instead.

My major goal for several years was to manage a 200-kilogram dead lift. I had increased the weight pretty consistently up to 190 kilograms, but at that point, it stopped. Every time I tried 200, it was as if the bar were cemented in the ground. And every time I lowered the weight in disappointment to 190 kilograms, I could suddenly lift it from the floor without a problem, sometimes even two and three times in a row! I stayed there for a long time (and also tried the weights in between, 195 and 197.5, with varying success).

One day at the gym I asked to take a turn when a guy was doing an endless dead-lift session. I counted to 180 kilograms on the bar and felt that would be enough for that day. I'd intended to lift it three or four times, but it felt so dead and heavy already on the first lift that I was content with that. When the other man later helped take off the weights, it turned out that the weight plates I thought were tens were actually twenties (the blue color they usually have on the edges had worn off so that they were black like the tens). So, it turned out I had lifted exactly 200 kilograms.

Micael

NUMBERS AND AGING

Age is just a number, it's said. And there actually is some truth to that expression. After all, the body doesn't know how old it is. According to the anatomy professor Leonard Hayflick, the body does not have *one* specific age; instead, it has several ages simultaneously. You see, the body consists of a lot of cells that divide and renew themselves at various rates. That rate differs between different body parts and organs in the body and between different people. The only thing cells actually have in common is how many times they can renew themselves, and Hayflick discovered that when a sufficient number of cells in the body have reached that maximum (called the "Hayflick limit"), then we die.

But even though we consist of all these different cells with their different renewal rates, most of us age at approximately the same tempo, year by year. The explanation is presumably, at least in part, that we use the same numbers—we count the number of years that have passed since we were born—to measure our age.

Unfortunately it's impossible, of course, to test whether that really is the case. To find out if we age at the same rate just because we measure our lives with the same numbers, we would need to compare people who measure their age in years with those who don't and see whether they age differently, which isn't possible. True, there are people who don't measure their age in years—for example, the Munduruku people in the Amazon, who can only count to 5—but paradoxically enough, it would be extremely difficult to determine how long they live based on all these cells that tick at various rates toward the Hayflick limit. Another way would be to fool people that they are older or younger than they actually are and see how they age, but no ethical review board would ever approve such an experiment.

Fortunately people are rather good at fooling themselves. We use the term "psychological age" for when we tell ourselves we're a different age than the calendar says. In several different studies researchers have compared psychological age with people's normal walking speed, and all the studies show the same thing: the lower someone's perceived psychological age, the faster they walk. The interesting thing about walking speed in particular is that it is often used as a simple epidemiological indicator of people's biological age and remaining life span. Walking speed is affected by everything from blood circulation and the respiratory system to muscles, joints, and the skeleton, and it works in that way as a summary of the total life force in the body, proven to be extremely accurate when researchers measured the length of life of hundreds of thousands of people. In other words, the slower you walk, the faster death will catch up with you.

Psychological age could of course influence how quickly we walk because those who actually *are* physically younger (and evidently walk faster) also *feel* younger because of that. But when researchers in one study compared walking speed in people with different psychological ages, they found a connection only when people had to report (and so be reminded of) an age number *before* they walked. The psychological age number therefore *led* to their walking at different rates. This also explains why researchers have found that trial participants guide their digital avatars in online gaming environments at a slower pace with the rising age of the avatars and actually even walk more slowly themselves when they then leave the experiment room.

Studies of tens of thousands of people show that psychological age, just like calendar age, influences everything we associate with aging, from memory and cognitive functions to physical health, feebleness, and mortality. How old you are and how long you live are therefore affected by the number you yourself put on your age. Literally.

That explains why our age has magic boundaries. Exactly as number boundaries influence how much we are able to lift, they influence how quickly we age too.

It's no secret that it hurts to get older. And that we humans easily end up in a little existential crisis when we pass a magic number like 30, 40, or 50. Especially men. Some buy a Harley-Davidson, a few enter into affairs, and others suddenly become exercise addicts.

My midlife crisis consisted of running a marathon for the first time. A hot and sunny day in Stockholm. And one of the first things I noticed among those who were at the starting line was how many other runners looked like me, slightly desperate 40- and 50-year-old men. One was even wearing a slightly undersized T-shirt with text that read, "50 and still hot."

After an incredibly painful run (never again!), I unearthed the statistic as soon as I had staggered back to the hotel. Sure enough: among marathon-running women and men, there is a rather dramatically large number of people who are just turning 30, 40, or 50. Among men the most common age for a marathon runner is 50, among women 30. And runners with milestone birthdays (30, 40, 50, 60) top the statistics regardless of gender. Statistics from over 2.3 million marathon runners show that people who are having a milestone birthday are overrepresented by a good 13.3 percent. That's quite a lot. Whether they're also in better shape is another question.

I'm not going to tell you how fast I ran. But what I can say is that the slightly overweight man with the tight "50 and still hot" T-shirt beat me by a slight margin in the sprint inside Stockholm's stadium. That was not exactly a boost to my ego.

Helge

Guess when people feel the oldest.

When they turn 42?

When they turn 40?

How do we actually react to passing a "magic" number like 40 or 50? Do we suddenly feel much older? Do we look at our bodies differently? Of all the strange things that have been researched, shouldn't someone have done research on this too? But no one has. So we did, inspired by the Stockholm Marathon.

We sent a questionnaire to several hundred randomly selected individuals of all ages, asking them their age, how old they actually felt (psychological age), and various questions about what kind of shape they were in. Then we compared all those who turned 30, 40, 50, and so forth with individuals of all other ages.

And what do you think we found? Well, for one thing, almost everyone feels younger than they actually are. Psychological age is consistently lower than actual age for people both "young" and "old." Maybe that's not so strange. People have a need to feel young and energetic. On average our psychological age is 8.4 years lower than our physical age, so we feel considerably younger than we really are.

But the funny thing is that individuals with an age ending in zero, who have just passed over 40 or 50, for example, on only that birthday feel relatively older than on other birthdays. If we subtract psychological age from actual age, there is a systematic difference between those with milestone birthdays and everyone else. Those with milestone birthdays feel on average only 6 years younger than their actual age—in other words, 2.4 years older than on all other birthdays. And when we ask them how old they feel their brain is, on average they feel that it is three years older than everyone else's.

> The week before my wife was going to turn 39, I asked her what she wanted to do on her birthday. "Yes, we probably have to have a big party anyway because I'm turning 40," she answered. When I pointed out that she was only turning 39, she raised her eyebrows and burst out laughing. "I thought I already was 39!" The next day she cancelled the eye examination she had planned in order to get glasses.
>
> Micael

Clearly, we are rather fixated on our age, and the number affects us. But is it right to calculate a human life in years? Would it perhaps have been smarter to count *days*? Would that influence our view of life in any way?

We tested that too. We asked 1,000 people to guess how long an average life is and gave them a number of alternatives, expressed in either days or years. Then we asked them how meaningful they felt that their lives were. Surely that shouldn't be influenced by whether they'd seen the alternatives in days or years, given that the length of time is the same, right? And besides, it's not really the length of time that's meaningful at all but what we do with it. As it turned out, it didn't matter—at least not for those who saw the alternative in words ("thirty thousand days," "eighty-five years"). Fascinatingly, however, those who saw the alternatives in numerals ("30,000 days," "85 years") thought that their lives did feel more meaningful when they saw their age expressed in days instead of years.

Our actual age influences us not only on milestone birthdays. We are reminded of the number (because it normally is a number—we are 21, 47, 69, or 85 years old) again and again in our everyday lives, and that has consequences.

We just described a study showing that being reminded of how old (or young) we feel can influence our walking speed. In that case, it was about psychological age, but what happens if people are reminded of their actual age? There was no such study, so we did our own. We asked over 2,000 individuals to do as many pushups as they were able to. Half of them had to write down their age afterward, the other half had to write down their age first—and therefore remind themselves of their age before they did their pushups. Not surprisingly the younger ones managed to do more pushups than the older folks: on average there was a difference of 25 percent between those below and above the median age—when they wrote down their age *afterward*, that is. When instead they wrote down their age *before* they did the push-ups, the difference was closer to 50 percent. Reminding participants of their age increased the difference between the younger and the older by almost twice (and the same applied to the difficulty they reported feeling).

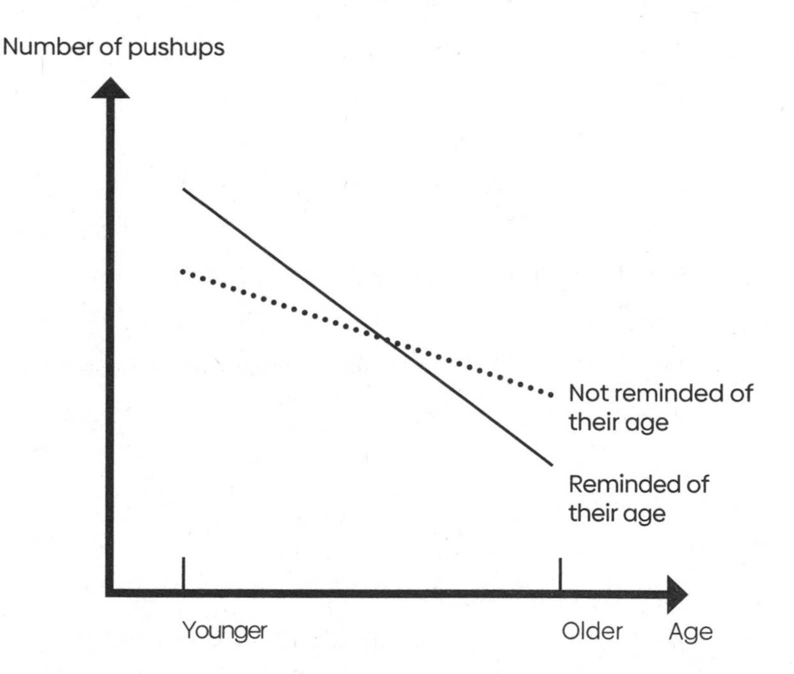

That numbers influence how we age and how many push-ups we do is both funny and worrying on an individual level, but on the societal level it can become rather unpleasant. Just think of all the times age comes up nowadays:

- Whenever you're asked something, whether it's for marketing research, polling, a public opinion survey, or the census, there's a good chance that you will also be asked about your age (and your answers examined in light of it).
- In your dating profile you are expected to provide your age, and when you view others' profiles, age is used as a filter.
- Whenever you take a medical exam, you are expected to report your age. The same goes for a lot of the health-related apps that track everything from your heart rate and pulse to your mood and your activity level.
- If you want to partake in a sporting event—for example, CrossFit Games or a race—you are likely to be asked to sign up for an age group.

It's not strange, then, that ageism is a growing problem in our societies where older people are discriminated against in everything from hiring processes to selection for sports teams, and that we risk both feeling—and becoming!—slower, weaker, and less on the ball when we see ourselves in our own and others' rising age numbers. With every milestone birthday, the gap between age groups seems to widen so that we become increasing foreign to one another and less inclined to associate. (Isn't it strange, by the way, that a 37-year-old is more inclined to reject a 41-year-old than a 31-year-old is to reject a 38-year-old in a dating app?) A large meta-study from 2020 of millions of people in 45 countries showed that ageism systematically reduces the opportunities for the elderly to get care and employment and leads to reduced interaction with other people, increased physical and mental ill health, and shortened life span.

BLAME IT ON THE SNARC

The influence of numbers on how strong or old our bodies are is an example of psychosomatics: the numbers get us to think (*psycho*) in different ways that influence the body (*soma*). But numbers influence us instinctively too before we even think about them.

Low numbers between 1 and 4 make us automatically move more easily to the left while high numbers between 6 and 9 instead make us more easily move to the right. There are quite a few amusing experiments that show this. For example, people have been asked to babble random numbers while they walk and then are suddenly told to turn in any direction. Those who said a low number immediately before were more inclined to turn left, and those who said a high number were more inclined to turn right. In the same way, those who just turned left were more inclined to say yet another low number while those who turned right chose a higher number. There is the same effect on how quickly people move in different directions, whether this concerns walking, running, or catching something that comes flying at you.

Our ability to catch something with our hands, regardless of whether it comes from the right or left, is also affected by numbers. You see, the muscles of the hand and fingers react automatically when we see and hear numbers. Lower numbers make the hands contract a little, and higher numbers make the hand instead open up a little; this has been tested by attaching electrodes to people's hands to measure the muscle activity, throwing things at people, and seeing if they're able to catch them.

Researchers call all these number-controlled movement patterns, which affect both our bodies and where we direct our gaze, SNARC, an acronym for Spatial-Numerical Association of Response Codes. Our inclination is to think of low numbers when we move backward and high numbers when we move forward. Likewise, we are faster moving (walking) upward when we think of high numbers and downward when we think of low numbers.

If you think about the numbers 1, 2, 3, and so on, you can surely see in your mind's eye how 1 is to the left and how 2 and 3 then continue to the right all the way to 10, like on a number line. Or how your inner eye wanders downward when you count from 10 to 1. There, incidentally, you have the explanation for why it's called "countdown" and why, in everyday terms, we talk about rising and falling numbers.

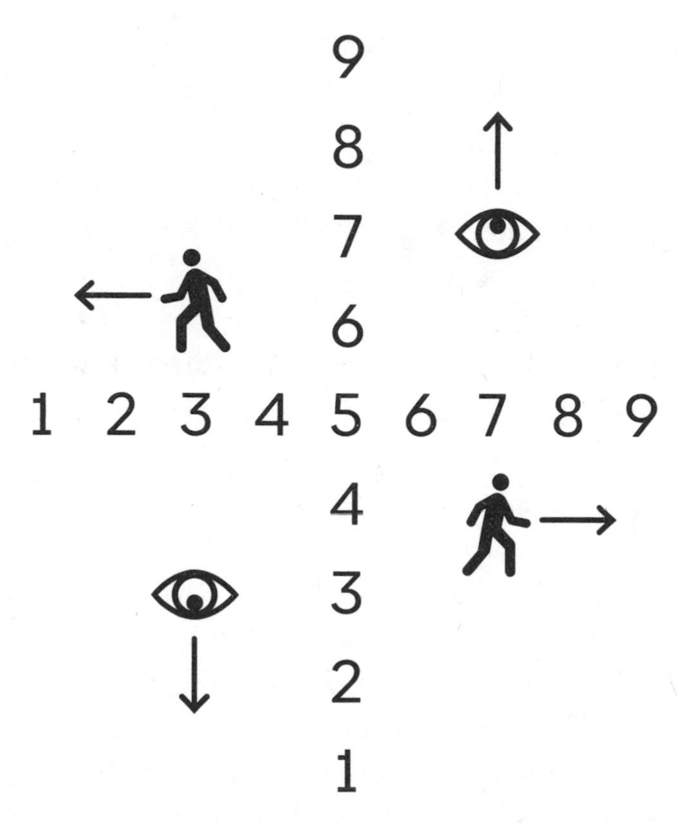

SNARC points to the fact that people's understanding of space and numbers is connected. And now it gets really interesting, because what connects space and numbers is a small part of the brain right behind the frontal lobe called the intraparietal sulcus (IPS). Brain scan studies show that the IPS is activated when we see and think about numbers. But also when we assess depth and distance. And when we direct our attention in different directions. And when we move and react reflexively with our hands.

Numbers are thus neurologically connected with many of our basal, physical actions, and we react to them more or less instinctively. Researchers sometimes call the brain cells in the IPS "number neurons"; they seem earmarked to react to numbers and do that in only a fraction of the time it takes for us to react to words.

Presumably this has to do with the fact that the brain connects numbers with our innate survival instincts. We are born with the ability to distinguish different quantities in the form of size and number among things around us. A four-day-old infant already seems to be able to see the difference between large and small blocks and between 1 and 2 blocks or between 2 and 3 blocks. At the age of six months, we start being able to translate numbers, so that babies who hear 3 drumbeats automatically look at a picture with 3 dots rather than a picture with 2 dots. They also react immediately when someone changes the number of blocks in front of them or when the size of the thing they see changes. Apes do that too, as do cats. Several amusing experiments have been done where apes and cats get to see 2 balls, whereupon the researchers place a screen in the way and take away or add a ball. When they then remove the screen, the apes and cats almost look shocked and seem not to believe their eyes. Reacting to number and size could mean the difference between life and death when it concerns quickly assessing enemies or access to shelter and food.

That would explain why humans are not alone in being able to count. It almost seems to be an animal instinct. Researchers have managed, for example, to teach both apes and cats, as well as pigeons and (of course, every lab's favorite) rats, to add together two objects before they get to eat them. Harvard researcher Irene Pepperberg even managed to teach her gray parrot Alex to count to 6 with several years of training. But while Alex and the other animals count the quantity of things (balls, seeds, peeps, words), we humans start at the age of four to distinguish ourselves from them by developing our unique ability to use and react equally to numerals. Long before most of us can read and write, we start learning numbers, connecting them with the fingers we count on, and creating particular number neurons in that part of the brain that guides the animal instincts to react to number and size. In other words we program the brain to react automatically and quickly to numbers as if it were a matter of life and death, regardless of what they concern.

In light of this, perhaps it's not so strange that numbers can influence us physically, making us stronger or weaker, younger or older, and get us to move in different directions. Or that numbers influence us in so many other ways, in so many other contexts, without our even really being aware of it. And that numbers are just as likely to upset as to help, because they connect instinctive, animal reactions with things they were never intended to be connected with.

ONE, TWO, THREE—MANY?

Numbers also make us sensitive to changes and differences that we might not notice otherwise. Take the Munduruku people in the Amazon, for example, who can only count to 5. Anthropologists visited them and asked them to solve problems such as choosing the larger or smaller of two piles of grain. As long as neither of the piles had more than 5 grains, there was no problem, but as soon as the piles had more than that, the Munduruku people had considerably more difficulty in choosing correctly. One pile had to be twice as large as the other, so that it was clearly visible which was largest, for them to be certain. The same thing happened when the researchers added or took away grains from a pile; when the pile contained more than 5 grains, the Munduruku people were no longer certain of what had changed.

The Pirahã people live in another part of the Amazon, and they have numbers for only 1 and 2. When anthropologists showed them a piece of paper with a number of lines on it and then asked them to draw the same number of lines, things went just fine as long as the task concerned only 1 or 2 lines. After that, it got harder, and only half could draw 5 lines.

In another experiment the researchers placed nuts in a tin can, angled the can so that the Pirahã participants could look into it and see how many nuts were there, and then set the can up so that they could no longer see inside it. They then took out 1 nut at a time and asked the participants to say when they thought the can was empty. When the can only contained 1 or 2 nuts, all the participants managed to guess correctly, but when it contained 5 nuts, only 4 of 19 managed to say when the can was empty. With 6 nuts only 1 in 10 answered correctly.

Even if there isn't a way to know whether the Munduruku people age more slowly than we do, we can probably be certain that neither they nor the Pirahã people suffer from the same kind of age anxiety, since even if they were to count their age in years, they would stop at approximately 5 and not be so sensitive to the differences after that. They probably wouldn't stress over likes on Instagram or become greedy for nuts either, because they haven't programmed their number neurons to count and see the differences.

Speaking of counting, what exactly was going on with Michael Jordan and his jersey numbers?

The leading explanation for why he made fewer points with the number 45 on his jersey is that he wore it after taking a one-year break from playing and was rusty; then he got warmed up again just as he changed back to number 23 and became the world's best basketball player again.

Some of the lessons from this chapter can function as a little number vaccine:

1. We are number animals and are influenced by numbers whether we're aware of it or not. For that reason, be careful with them, for both your own and others' sakes.

2. Take a moment to think when you react instinctively to numbers. What do they actually mean? (Nowadays, it's seldom a question of survival, food for the day, or friends and enemies.)

3. Remind yourself that magic boundaries don't really exist; the difference between 39 and 40 is the same as between 38 and 39 or between 33 and 34.

4. Don't let numbers determine your age, your strength, or who you are—because they will if you let them. Decide your numbers for yourself instead.

5. The next time you play basketball, choose a jersey with as low a number as possible so that your opponents will instinctively pull a little to the left, and then dribble to the right. It should work.

With this little number vaccine, we hope you can become aware of and get better at handling the effects of numbers on your own and others' physical self. Speaking of awareness—have you thought about how numbers influence our psychological selves?

3

NUMBERS AND SELF-IMAGE

On April 17, 2020, 18-year-old Noor Iqbal and his father Parvez were having lunch together in their house in Noida outside New Delhi in India. Parvez went out to buy vegetables and came home to a locked door, bolted from the inside. With the help of the police, he broke down the door and found Noor dead. The police investigation came to the conclusion that the teenager had died by suicide, apparently after becoming depressed by getting too few likes on his TikTok videos.

Sadly, Noor's case is not unique. Chloe Davidson of Lanchester, England, was 19 years old and wanted to be a photo model when in December 2019 she took her own life, affected at least in part by too few likes on the pictures she posted on social media. There are a number of similar examples, and even if they are fortunately rather rare, they illustrate what, in the worst case, can happen when judgments from other people become extremely visible, public, and measurable.

Suicide accounts for 13 percent of all deaths among teenagers in the United States, and the new currency of social media in the form of likes, hearts, shares, hits, and followers may play a role in many of those cases. A low number of likes alone is unlikely to kill a person, but such an extreme quantification of a person's popularity and worth can reinforce existing psychological and social mechanisms. The number of likes becomes an amplifier of both fragile and inflated self-images and, in a few hours or minutes, can destroy or pump up an ego. Being exposed to your own numbers means that the weak feel weaker and the strong, stronger.

On Facebook, still the world's biggest social network, more than five billion likes per day are generated, corresponding to four million likes per minute. On Instagram we like almost two million images of friends and acquaintances every minute. All these images and posts are clearly marked with the number of likes so that anyone can see how popular or unpopular we and our vacations, children, hobbies, dinners, and beach bodies are.

But what do these figures have to do with self-image and self-confidence? And what about all the *other* numbers that we're constantly fed with in places other than social media? What does knowing the balance on your checking account, the number of bonus points you've earned, your pulse, and the number of steps you've taken today do to you? Through various apps and digital interfaces, we are fed around the clock with information about our own numbers, our own achievements, and our own measured performance. Does this influence our self-image and identity more than we think?

PULSE AND MONEY

A very long time ago, before the internet, smartphones, and all the other connected devices, we had fewer numbers and quantifiable units to measure ourselves and one another with. We probably knew more or less how old others were, how many children they had, and the number of arms and legs they had, but most other characteristics we had to estimate, guess, or discuss with others to find out. In the same way we also had fewer cold facts about ourselves. We didn't know how many people liked our cat, how many steps we'd taken in a given week, or exactly how many colleagues actually enjoyed the column we wrote in the newspaper. We were living in self-quantified darkness.

There was nonetheless *one* very helpful number—an important scale that everyone adhered to and with which we measured both our neighbors and ourselves. An important, quantifiable, socially visible, and tangible unit that has been with us for centuries: money.

Money has always been easy to compare, simple to measure, and very important to people. Money has always given status, self-confidence, and social capital and throughout history could be compared quantitatively between all kinds of people—not all that different from "likes" in social media. For just that reason it may be interesting to look a little closer at the research linked to the psychological effects of money when we're trying to understand what *other* quantitative units and self-referencing numbers do to us.

Merely looking at or thinking about money influences us in more ways than we think. Simply seeing a picture of money, picking up currency, or even touching pretend money does something to people's thoughts and behavior. Decades of research on the effect of money—looking at behaviors in people who were reminded of money versus in people who weren't—clearly show that it makes us more focused on ourselves, that it makes us feel stronger and more self-confident. Individuals who are exposed to money feel that they have more control over their own lives, are more independent, and need other people to a lesser degree. Studies even show that money makes us less afraid of death. People who get to see and pick up money, whether real or pretend, feel less fear of death than those who have not been exposed to money.

When we get to see and handle money, we can also become

less willing to help others, adopt a more transactional view of the world, and become more insensitive. Randomly selected individuals who are exposed to money show less consideration and are less social than everyone else but feel independent and have plenty of self-confidence where making things happen is concerned. Not particularly charming, perhaps? Some call this the "asshole effect," inspired by the stereotypical behavior of rich people who think they own the world. The fascinating thing nonetheless is that this concerns all of us, not just people who *have* money. When quite ordinary, randomly selected individuals are reminded of the concept of money, the effects are the same. They become more calculating and focused on themselves and have greater self-confidence.

Okay, so what can this tell us about the effect of all the other numbers and quantitative units that we encounter daily? Can it be the case that the number of followers and bonus points or the numbers on a Fitbit do similar things with self-confidence and self-image? What do you think?

A simple way to find this out is to take a look into the brains of people who use social media. Curious as we are, we did a study of over 300 Americans with Instagram accounts to see if there was any connection between the number of likes and self-confidence. And hardly surprisingly, the number of likes fit rather well with both self-confidence and self-sufficiency—that is, how well you think you manage on your own. In the study the number of likes on every Instagram picture was 15 on average. When we then looked more closely at the test subjects who on average had *fewer* and *more* likes per image, we saw a fascinating pattern: the result concerning self-confidence, general satisfaction with life, and independence was considerably higher for those with many likes than for those with fewer likes on their pictures. Those with many likes also reported lower stress levels.

Of course it may be the case that individuals with poor self-confidence, high stress levels, and low satisfaction with life take lousy, boring pictures that no one likes. Or that they have very few friends. It doesn't seem a likely explanation, but we can't rule it out. To be able to say anything about cause and effect, we must also experiment a little to see if higher numbers really do increase people's self-confidence and mean that they feel stronger and better.

For that reason we did two experiments. The first was on a group of exercisers in the United States. To investigate if higher and better numbers also lead to increased self-confidence outside contexts such as money and social media, we focused on numbers that concerned how fast they ran. And to achieve that we had to fool them a little. We divided the exercisers into three random groups and told one-third of them that they ran faster than the average and one-third that they ran slower than the average. The final third was our control group; they got no information at all about how fast they ran relative to others. There were, of course, no major differences among the three groups concerning how fast they *actually* ran; we just played a little with their minds.

And what happened? Well, those who were told that they ran faster than the average reported generally higher satisfaction with life, greater self-confidence, and lower stress levels than both the control group and those who thought that they ran slower than the average. The poor things who thought they ran slower than average suddenly felt that life was heavy and difficult and found that they were able to manage on their own to a lesser degree, despite the fact that on average they were in just as good shape as the participants in the two other groups. Even more fascinatingly, we also measured how physically attractive the participants in the study considered themselves to be. Those who thought they ran faster than the others suddenly thought that on average they were more physically attractive; those who erroneously thought they ran slower suddenly thought that they were a little uglier than average.

Our second experiment was on 400 users of Instagram, also in the United States. The investigation was very simple and conducted online. First, we asked the participants about their age and gender, as well as how many followers they had on Instagram. We then explained that "our algorithm" calculated how many more or fewer followers they had than others with the same demographics, but here, we must admit, we fooled them a little. We had no such algorithm but instead randomly divided the participants into two groups. One group was told that they had 39 percent *more* followers than others in their demographic group, and the other group was told that they had 39 percent *fewer* followers.

You can probably guess what we found out. Yes, having more followers led to better reported self-confidence and greater satisfaction with life compared with those who (believed they) had fewer followers. And keep in mind: the participants were randomly divided into these two groups, so there was no difference between them from the start. The only thing that changed was that these individuals suddenly *believed* that they had more or fewer followers on Instagram than others like them.

By the way, guess what happened when we asked them to choose between two prizes they could win as thanks for

participating in the experiment? The first choice was a solo cooking class with a master chef; the other was a cooking class together with friends. Well, those who heard that they had more followers than average were more likely to choose the elite, solo cooking class—an effect not completely unlike what money does to us.

DOPE AND DOPAMINE

Studies have shown that getting many likes on social media gives the brain a dopamine kick. In 2016 a brain scan (fMRI) was performed on a group of American teenagers while they looked at pictures in an app that resembled Instagram. The pictures were both the participants' own and strangers', and the researchers (randomly) gave each picture a varying number of likes. When the teenagers saw their own pictures with a high number of likes, the researchers observed increased activity in major parts of the brain. The greatest increase was in those parts of the brain that are linked to reward but also in the area called the social brain and the area connected to visual attention. The researchers drew the conclusion that likes contribute to making social media addictive and influence the brain in a similar way as gambling.

There is no shortage of research that identifies the problems with quantification and likes in social media. A number of studies point out that dependence, narcissism, and depression can develop with use of social media. The connection between the number of likes and people's self-confidence is also well documented. The more likes, the better your self-confidence. The fewer likes, the worse your self-confidence. One of the reasons that the number of likes has such a direct and immediate effect on self-confidence is that they make social comparison so incredibly simple. Two numbers are extremely easy to compare. Two vacation pictures or two pictures of a plate of food are not. Pictures give space for interpretation and ambiguity and are subjective. You may think that your vacation is just as good as mine if you only look at two, often extremely different pictures. But if your vacation picture gets 200 likes and mine only 50, it appears obvious to everyone, even to me, that your vacation is better than mine.

The paradoxical thing with the whole mechanism about likes and self-confidence is that it seems to do damage at *both* ends of the scale. Those who don't get as many likes risk becoming depressed and have worse self-confidence. Those who get lots of likes risk becoming self-absorbed narcissists.

COMPARISON HELL

We humans simply *love* to compare ourselves with others. In order to understand the world around us we need to know whether others are similar, better, or worse. When we meet new people, we quickly judge whether they're higher or lower than us in the social order or hierarchy, and we love to rank and categorize one another based on every conceivable aspect. For that reason we have ranking lists for almost everything. We have sports standings, hotel ratings, global happiness measurements, credit ratings, and lists of the best schools, hospitals, and airports.

The main reason that social comparison increases our performance and motivation is that an unfavorable comparison is perceived as a threat to our own self-image—and that motivates us to perform better next time. If other people run faster or get more likes on Instagram, you want to reach their level, or ideally outclass them. And the more important the comparison is to you, the more motivated you are to improve. Neuropsychological studies show that social comparison is closely linked to the reward center in the brain. If you perform better than others, you feel happy; if you perform worse, you feel sad or angry.

Do you know, by the way, who is happier: the person who wins the silver or the bronze medal at the Olympics? There's research on this too. It's possible to interpret and code facial expressions among athletes both when they reach the finish line and when they are standing on the podium when the prizes are awarded. And what do they find? Well, bronze medal winners are systematically much more content than silver medalists. But how can this be when the bronze medalists performed worse? Well, the silver medalists "lost" the gold, while the bronze medalists won a place on the podium. The silver medalists compare themselves with the gold medalists, while the bronze medalists compare themselves with those who didn't win any medal at all.

We humans compare ourselves socially to a higher degree "upward" than "downward," which you might intuitively think is a good thing. Comparing yourself with people who are more capable, faster, and smarter gives both inspiration and motivation to improve. Unfortunately, the exact opposite happens. Social comparison upward makes us dissatisfied with ourselves to a greater extent. Just ask all the silver medalists in the Olympics.

Or look at what happens on Facebook and other social media. There, the users themselves can choose if they want to look at, follow, and compare "upward," that is, with individuals who are better looking, richer, and have more likes and followers, or "downward," with individuals who appear to have a rather miserable life. And surely you know what most people prefer to do. Yep, they compare themselves to a greater extent with individuals who have more likes, more followers, more friends, and higher numbers. Studies show that this not only makes them less content with their own lives but also that they overestimate how good others have it. And that effect is strengthened by the fact that people post beautified pictures from their lives on social media—not pictures that show what life is *really* like. In an experiment where individuals had to judge different (fake) profiles on social media, the researchers found that the choice to socially compare "downward" had no effect whatsoever on the individuals' self-confidence. You might think that it would give these individuals a self-confidence boost, but it didn't. However, when the individuals compared "upward," their self-confidence and their assessment of their own lives went down. For the ego, then, it seems to be a clear disadvantage to open a social media app and scroll through other people's likes and profiles.

It's actually similar to watching TV. People on TV and in TV series are generally a tad richer and more successful than the average population. What do you think happens with people who watch a lot of TV? They believe on average that other people around them are richer than they actually are. They also underestimate their own wealth and happiness.

And what do you think happens to people's motivation and contentment when they find out how much their colleagues earn? They find new things in life to be dissatisfied with. There is always at least one colleague who earns an amount we deem to be undeserved, isn't there? In a study of 5,000 British employees, researchers found that the more colleagues earned compared to themselves, the unhappier they were. In another study, conducted on students and staff at the prestigious Harvard University, half of the respondents said that they would rather earn $50,000 if their colleagues earned $25,000 than earn $100,000 if their colleagues earned $250,000. It's rather startling that people would rather cut their salary in half than be the lowest-paid person at work.

We see social comparison everywhere, and we compare both consciously and unconsciously all the time. And so it's not just the number of likes, views, shares, and followers that form our yardstick. The same thing applies to all of the other numbers in our lives: income, weight, height, steps per day, average walking speed, bonus points, levels in various games—everything. New sensors, increased digitalization, and globalization also mean that we get more numbers all the time, about ourselves and those around us. We get a yardstick on everything. And what is perhaps even more troubling is that the sorts of things that in the past were totally impossible to compare are now compared easily. Before, there were small refuges that the numbers hadn't yet managed to infiltrate, where we ourselves were forced to think, reason, and assess and subjectivity was essential. Where my personal assessment of something was just as correct or crazy as yours. Where things couldn't be constantly compared with each other. Those days are gone, my friend.

The incomparable has become comparable. Everything can be reduced to a number and a yardstick.

Do you wonder if you're too fat or too thin? BMI will give you the answer. Do you wonder how physically attractive you are? The number of likes on selfies or swipes on Tinder will give you a clear response. Are you financially responsible? Check your credit rating. Did your neighbor have a nicer holiday than you? Just have a look at their hotel on Tripadvisor.

The companies that compile this data about our lives have naturally cottoned on to the negative psychological effects this can have. The first five years that Facebook existed there was no "like" button on the network service. But since it arrived that little button has acquired great significance for both Facebook's and other social media's spread and commercial success. In pace with the fact that research has documented a number of negative psychological effects connected to likes and quantification on social media, the social media companies have also come under pressure to do something about it, which was why, in 2019, Instagram (itself owned by Facebook) tested a change in the service whereby users could like pictures but not see the number of likes that other users had received or how many times a video had been seen. The test was conducted in Canada and after that in six other countries. "We are testing this because we want your followers to focus on the pictures and videos you share, not how many likes they get," representatives for Instagram commented, but the challenge for the company naturally is that such a change can reduce how attractive the service is, how addicted the users become on it, and, therefore, how many times they click on it a day. The direct response among users when the test became known was also very negative, and many felt that Instagram introduced a change that "no one wanted." The test results and conclusions are still not known, and Instagram users can still see both the number of likes and how many times a video has been downloaded.

> My daughter told me that everyone (?) has "finstas" on Instagram, that is, "fake Instas," extra Instagram accounts alongside their own more "official" account. Some have finstas to be able to share less perfect and unedited pictures with their very closest friends, but the majority create such "fake" accounts to be able to like their own posts. "You want to get as many likes as possible." I asked whether this is to impress others, and she shrugged and said, "It just feels better." That reminded me of an expression I learned in the United States when Instagram was new: "instacurity." That's when people get anxious after they just posted something and the likes are not pouring in.
>
> Micael

I AM A TRAVELER

That the numbers around us influence both self-confidence and how we feel is indisputable. But do they also influence our identity and what interests us?

Everyone who has had a job in either the private or public sector knows that the numbers you are measured by become important. Often a little *too* important. Whether you measure customer satisfaction, profitability, or sales, the numbers slip into your head and affect motivation, choice, and prioritization. In our work as professors at two Nordic schools of economics we are constantly being measured from every conceivable angle: instruction evaluations, number of mentions in the media, number of scientific articles, the "impact factor" of those articles and journals, number of citations, h-index on Google Scholar, ResearchGate score, and dozens of other measurement parameters. And just because the numbers are so visible, simple to compare, and supposedly objective, they become important—for employers, for colleagues, and for how we see and assess ourselves.

But the numbers aren't just at work. Take a quick look at the apps you have on your smartphone. Then think about what numbers the apps are feeding you with and reminding you of. Depending on your interests and your personality, you presumably get numbers about most things in your life. You are fed with information about your financial situation: bank account, loans, creditworthiness, pension, funds, and stocks. In addition you get information about your health: number of steps, miles walked, pulse, average walking speed, and elevation gain. You get numbers from social media about the number of views, likes, followers, shares, and hits. And you get information about member points here and there, what level you're at in *Candy Crush* and *Hay Day*, your energy consumption, your rankings on Tripadvisor, Airbnb, and Uber, and a long series of other numbers from employers, apps, and sensors.

And because those numbers seem objective, true, concrete, clear, universal, and easy to compare, they become important and begin to influence what you focus on, how you prioritize, and perhaps also how you look at yourself.

> "You are a true traveler," the SAS (Scandinavian Airlines) app tells me. The evidence is clear: 6.7 times around the world and 504 hours in the air since 2003. 213,726 EuroBonus points on the account. And even if I don't feel like a real globetrotter when the numbers are presented, the more often I open the SAS app, thereby exposing myself to the numbers, the more I integrate them and what they stand for into my self-image. I am a globetrotter. A real international man of travel. That is what I am.
>
> Helge

The numbers on your smartphone and what they represent subtly influence your identity; they have an insidious, self-reinforcing effect. If you suddenly get a lot of shares on Twitter, you consider yourself an important public debater, and you presumably tweet even more. If you get a lot of likes on your exercise pictures, you post more. Exercise becomes successively more important for you, and a greater share of your posts on Instagram contain exercise photos, because lots of likes release lots of dopamine, strengthen your self-image, and mean that picture, that activity, or those clothes become more important to you.

So if everything that can be measured becomes more important to your identity, and if the numbers influence self-confidence and self-image to such a high degree, then perhaps you should be a little more attentive to what numbers you handle in your everyday life.

Here are five small pieces of number vaccine advice for your self-image:

1. Be attentive to the fact that numbers and money have a lot in common. They can make you more calculating, egotistical, and self-absorbed. And you don't really want that, do you?

2. Keep in mind that *both* low and high numbers can destroy your self-image. The low ones can sabotage your self-confidence; the high ones can make you self-absorbed and narcissistic.

3. Numbers, especially in social media, can be addictive. Do a detox now and then!

4. Keep in mind that experiences are subjective. Two runs, vacations, or meals can't be compared.

5. Don't let the numbers control who you are. Remove the kind of numbers on your screens that lead you away from who you are and want to be.

With these tips, we hope that it's going to feel a little easier and better to just be yourself, regardless of what the numbers say.

NUMBERS AND PERFORMANCE

In December 2010 the business angel, health freak, and author Timothy Ferriss proudly presented his new book, *The 4-Hour Body*, with a big, fresh smile. If you can believe the subtitle, the book is "an uncommon guide to rapid fat-loss, incredible sex, and becoming superhuman." It quickly climbed the *New York Times* bestseller list and inspired a new generation of self-trackers with new tips and methods for improving their lives. Ferriss shows readers how, through precise monitoring of their weight, health, sleep pattern, and much else, they can achieve increased performance and, like Ferriss himself, gradually become superhuman. The tips include how to manage on just two hours of sleep, achieve fifteen-minute orgasms (if you're a woman), increase fat loss by 300 percent, triple your testosterone levels, and repair permanent physical injuries.

Timothy Ferriss, who has gotten filthy rich from all his podcasts, books, and consulting for Uber, Facebook, Shopify, and Alibaba, is an enthusiastic practitioner of self-monitoring, an adherent of the "Quantified Self" movement. Ferriss not only observes his own heart waves when he's sleeping but has also had a blood sugar meter surgically implanted in his stomach in order to get real-time data on his glucose levels. He has also had a biopsy done on his thigh to measure enzymes and muscle fibers. The number of apps, sensors, and monitoring apparatuses in his life could easily make NASA appear technically obsolete.

Timothy Ferriss calls all this scientific self-experimentation, while others might perhaps characterize it as deep navel-gazing or intense self-absorption. But Ferriss is far from alone. Studies show that almost half of us register one or more flows of health data about ourselves. Sales of Fitbits, Apple Watches, and various sensors have gone sky-high. And the Quantified Self movement now has members in over 34 countries divided among over 100 local chapters. The biggest groups are in San Francisco, New York, London, and Boston. There is even a separate division in the movement called the "Quantified Baby," whose members use various sensors and software to collect data about their babies' daily activities and health.

How on earth did we end up here?

As we know, the fascination we humans feel for numbers and data about ourselves is not a new phenomenon. Nor is self-quantification. The Pythagoreans did it over 2,600 years ago. We have probably felt an urge for numbers about ourselves ever since the beginning of time. If, for example, Benjamin Franklin were alive today, he would presumably be a very enthusiastic life blogger with hundreds of thousands of followers and his own podcast. Besides being a Founding Father of the United States, a musician, an author, and the inventor of, among other things, the lightning rod, Franklin kept an incredibly detailed diary packed full of numbers about himself and the life around him. He used the diary and the numbers as a basis for self-reflection and self-improvement and focused on 13 virtues that he daily followed and monitored. The Quantified Self movement considers Benjamin Franklin its ancestor, and on devoted self-quantifiers' websites, you can find his words of advice for productivity and division of the day's hours into various events and work tasks. Philosophers such as Michel Foucault, who emphasized the importance of thorough knowledge of oneself in order to develop and improve as a person, are also considered part of the ideological framework behind the self-quantification movement.

THINNER, HEALTHIER, FASTER?

Today, when people can acquire smart watches, smartphones, and countless logging apps, we have opportunities for self-quantification that Benjamin Franklin could only dream of. Self-logging has become an everyday sport, and there are a great number of books and websites about the subject, as well as hundreds of apps. We log and monitor ourselves in order to get thinner, healthier, and happier, to be able to run faster and perform better. Over 40 percent of all Americans think that self-monitoring increases athletic ability and reduces fat.

So a natural question is, Does it work? Do we really get thinner, healthier, and happier from this continuous monitoring of numbers about ourselves?

The research appears somewhat divided. The majority of the (few) controlled studies that look at the effect of smart watches, step counters, and various forms of logging of health data find a significant but relatively weak positive impact on the person's health and performance—regardless of whether it concerns weight loss or exercise frequency, intensity, or performance. If we use a Fitbit, Apple Watch, or other way of monitoring our own health or performance, then we run a *little* faster, lose a *little* more weight, or perform a *little* better. But just a little. There are also relatively large individual differences between different people. It works for some but not for everyone. And there are indications that the effect is relatively short-term and temporary.

How can this be?

Researcher Jordan Etkin at Duke University has conducted a series of fascinating studies dealing with self-quantification, performance, and motivation. In one study she had people perform various positive activities, such as exercising or reading a book. Half of the participants in the experiment were told about their own performance in numbers (how far they'd walked or how many pages they'd read), while the rest were not. Afterward she measured the participants' performance, motivation, and happiness levels. She also investigated whether the participants chose to continue with the activity after the experiment was concluded. And what did she find? Well, as in many other studies, it turned out that this business of monitoring and quantifying your own behavior led to *slightly* better performance. The participants who were given numbers about how they performed walked a little faster and a little longer or read a little more. But their motivation was reduced, and they continued with the activity to a lesser extent after the experiment was over. Self-quantification meant that over time the participants liked the activity less and cut down on it. Those who logged their own performance also got lower scores on satisfaction and happiness than those who had done the exact same activity but were not measured and quantified. The results were the same whether Etkin forced the participants to self-quantify or they themselves chose to do so voluntarily.

Why does this happen? The process of measuring forces us to be more attentive to what it is we are measuring. If you measure how many steps you take, you'll become more focused on that. If you count the number of pages, you become more focused on how much you've read. And even if you yourself don't explicitly want to walk farther or faster, we know from the research that the measurement in itself means that people will get better. When you measure heart rate, speed, and distance on a jog, little by little you focus on these numbers more than on the reason you wanted to jog to start with. By focusing on measurement and extrinsic motivation, you start doing once positive, fun activities more for the sake of their usefulness than for enjoyment. If you are someone who likes to jog in order to get fresh air, listen to good music, and experience nature, the intrinsic motivation connected with this is gradually replaced by performance, work, and extrinsic motivation the moment you connect to Fitbit or Strava.

A lot of fun and parallel research on children also supports this. For example, preschool children who were told that they should eat carrots because then they'll get good at counting eat fewer carrots and think they taste more disgusting than other children do. And if you reward a child for drawing, the child soon finds drawing boring. When driven by extrinsic instead of intrinsic motivation, an activity becomes less attractive and fun. Eating carrots becomes blah, jogging becomes a task, and reading becomes an effort.

Living evidence that self-quantification can go terribly wrong is the Norwegian Torbjørn Høstmark Borge. Borge liked exercising and gradually started using an activity meter and Strava. He really regrets that today. "I noticed that every time I activated the step counter I got completely manic," he told the newspaper *Bergens Tidende* in September 2020. "You get constantly pushed to achieve new goals. That has hung over me as a pressure every day and uses up a great deal of energy and focus." Eventually, his Strava dependency led to rhabdomyolysis (deterioration of large quantities of muscle cells) after he pushed himself to always exceed 40,000 steps a day. "It started with at least 20,000 steps a day, and suddenly it was never below 35,000. Then it was the same thing with 40,000." Both Borge's joy in exercising and his body were destroyed by extrinsic motivation and counting.

We humans often use extrinsic motivation to get *others* to perform better. Parents reward their children with ice cream and chocolate, and companies motivate their employees with money and bonuses. Sometimes this works, at least in the short term. But as we've seen, extrinsic motivation very quickly consumes our inner motivation. If you get paid for performing an activity that you love, you risk its feeling like a burden after a while.

In that way Etkin's research is reminiscent of all the research dealing with money and motivation. Here, however, the extrinsic motivator is not money but the number of steps, likes, or pages read. We know that money can get you to perform, but you soon get tired of the activity you're paid for because, over time, you connect your effort to the reward, not to your own inner motivation. In the same way the numbers you choose to produce about yourself—average walking speed, number of steps, number of likes, and bonus points—can gradually reduce your own internal motivation.

YOUR HEART, YOUR DATA?

If you happen to be a doctor, civil engineer, auditor, or one of the very many people who eagerly register their own health data or exercise performance, perhaps you think that the previous sections are overly negative and paint a bleak picture. Surely it can't be terrible to keep track here and there, you think. You have your Fitbit use under control. You smile when you're jogging. And you're actually content with your data.

Besides, maybe the numbers you register truly are meaningful for your health. If you're overweight or have high blood pressure, it's good to keep an eye on several numbers and data flows from the body. And if you have a condition such as diabetes, it's very important that you keep track of the glucose levels in your blood, which small, subcutaneous sensors can now effectively enable. Accessing data and numbers about your body and health can be incredibly valuable, or even necessary, for some people—just ask Hugo Campos, whose story is told by the Medicine X project at Stanford Medicine.

For his entire life Hugo Campos felt that his heart behaved strangely. He got palpitations. His heart skipped a beat. *Too much coffee*, he thought. Or not enough sleep. One morning in 2004, when he was running to catch the subway, he became nauseous and fainted. After many investigations, doctors at Stanford determined that he suffered from hypertrophic cardiomyopathy, a serious condition that causes the walls of the heart chambers to thicken. Three years later, in 2007, a defibrillator was surgically implanted to monitor his heart rhythm. All the data from the defibrillator was streamed to the manufacturer Medtronic, which forwarded it to his doctor. Campos, who had lived with an irregular heart rhythm his whole life, therefore looked forward to getting insight about the data from his defibrillator. But in 2012 Campos lost his health insurance. Without access to doctors and thus to his own data, he took matters into his own hands. On eBay he found a device that could be used to reprogram the defibrillator, and then he tried to hack his way into it. In order to experiment and get more knowledge about the defibrillator in his body, he even visited a funeral home, which sold used defibrillators that had been removed from bodies before they were cremated. But this attempt also proved difficult. The manufacturers of the defibrillator had become sensitive to data security after an incident in 2011 when researchers at a conference hacked their way into and took control of a defibrillator in real time while onstage.

Since 2007 Hugo Campos has worked for a change in the regulations and technology companies' practice so that patients can get access to their own health data. But in 2022, 15 years after the operation, Campos still did not have access to the data that his $30,000 defibrillator collects about his heart and body and forwards to a cloud service.

Campos thinks that if he'd gotten access to the data flows from his defibrillator, he could simply have connected this data to the flows from his Fitbit and to many other activities in different aspects of his life—for example, to find out how his heart rhythm is affected by coffee, alcohol, certain medications, and various forms of exercise. Campos thinks that he himself is better equipped to register and experiment with this than doctors whom he sees only sporadically and who don't live with the disease every day. Like patients who have diabetes or other conditions that require monitoring, Campos believes that access to his own health data is his right and important in principle.

Here is a considerable paradox. We have access to large quantities of numbers about our own bodies and health that perhaps increase our performance a tad but over time often kill motivation and joy. However, the most critical numbers, those that could potentially radically improve quality of life and coping for people like Campos, are owned by pharmaceutical and technology companies.

BIG BROTHER

Have you thought about who actually owns and has access to your health data? After Google reported in 2019 that it would buy Fitbit for $2.1 billion, several IT experts and ordinary consumers stopped using their devices because they didn't want to give Google access to data about their sleeping patterns, exercise, and health. *Google already knows enough about us*, they thought. Gradually more and more people became skeptical about the massive purchase of health data, and in August 2020 the EU Commission announced it would undertake a full-scale review of the purchase and of Google's access to people's health data. Fitbit has sold over 100 million units and has 28 million active users, so that adds up to a lot of jogging sessions, heartbeats, and location data. Google argues that through the purchase of Fitbit and increased use of artificial intelligence, it will be able to give people even more and better data about themselves so that they can learn more, feel increased self-awareness, and improve their lives. This is the future. Thanks to more sensors both on and in the body, on phones, in bed, at the workplace, at home, and in the car, we can all perform (*a little*) better.

We don't just think that numbers, measurements, and comparisons make us better, faster, and more effective on an individual level. This belief permeates everything from the reward system and key ratios in our companies to grading systems in school, child-care standardizations and measurements, and internal price systems in health care. And because we perceive numbers as exact, universal, eternal, and comparable, we see decisions and systems based on numbers as objective and transparent. Of course, they aren't—that's nonsense—but the alternative to numbers is clearly worse. What would we measure instead?

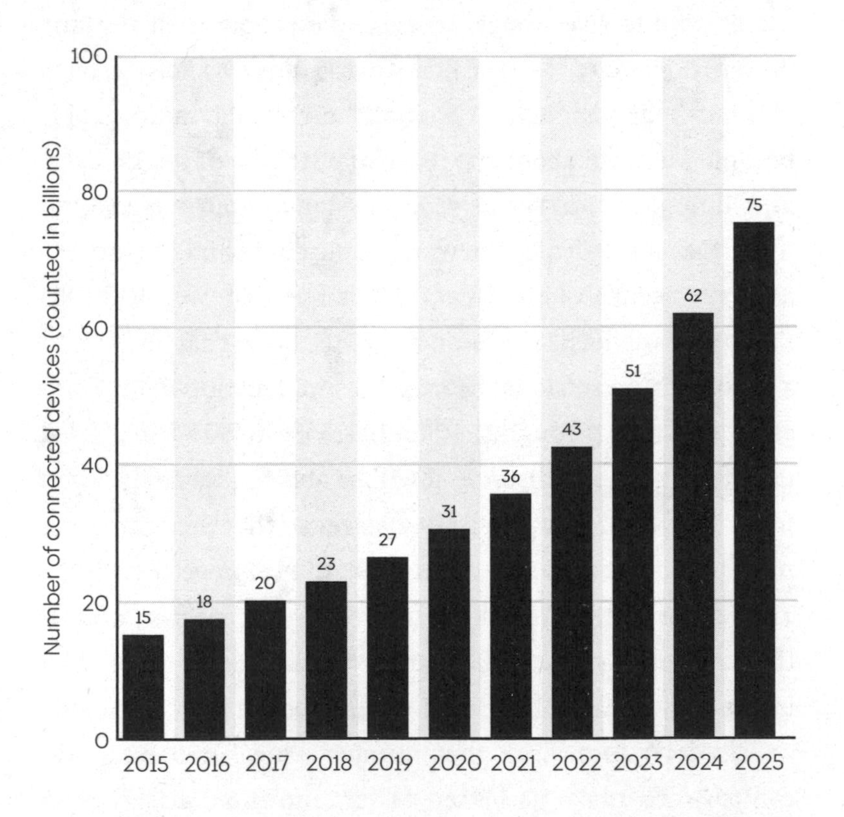

Interestingly, people in countries like Sweden and Norway don't seem to believe as much in the power of numbers and measurements when it comes to getting us to perform better. Sweden and Norway are countries with a high degree of trust, both among citizens and between them and the public sector. There the role of numbers in the public systems and politics is less prominent than in countries with a lower degree of trust. In the United States, in contrast, there has been a culture of fundamental public distrust of the public sector ever since the 1960s. And what is the result of that? Numbers and measuring systems replace subjective assessments and experience-based decisions everywhere from schools to police departments. Things become extremely rigid and not necessary especially effective. *Computer says no.*

But even in totalitarian regimes, faith in numbers as performance enhancers is strong. China's social-ranking system is perhaps the most extreme example. The system, rolled out in 2020, contains a series of databases and monitoring systems that assess whether individuals, companies, and organizations can be "trusted." Every individual gets a score, with rewards for those whose score is high and penalties for anyone with a low score. If you have a low score, you're likely to have to deal with restrictions concerning education and travel, slower broadband, and higher interest rates on loans. As China's prime minister Li Keqiang explained in a speech in 2018, "Those who lose credibility [that is, get a lower score] will find it hard to make a tiny step in society." The "carrots" used to promote citizens' performance are all the advantages you get if you achieve a good score. They may take the form of prioritized health care, reduced tax, and better financial terms and credit. Data is obtained from both traditional sources, such as crime registers, public authorities, and economic registers, and from third parties, such as online credit suppliers. Chinese authorities are also experimenting with automated data capture via video and internet monitoring.

The pilot version of China's social ranking system has already produced results. During 2018 citizens with low scores got to feel various restrictions.

128 persons were not allowed to leave China because of unpaid taxes

Citizens were not allowed to take management jobs or to represent a company in legal questions **290,000 times**

1,400 dog owners lost points, were fined, or their dogs were taken away because they didn't pick up dog poop or they let their dog be unleashed

Passengers were denied purchase of train tickets **5.5 million times**

Passengers were denied purchase of airline tickets **17.5 million times**

Source: Visualcapitalist.com

Faith in numbers and measurement as something performance enhancing and possibly disciplinary is thus central in most cultures, whether you find yourself in the United States, Norway, or China. Perhaps with the exception of the Munduruku and Pirahã tribes in the Amazon, people and societies all over the world are affected by the number epidemic, believing that through numbers you can monitor, stimulate, and motivate as well as increase performance.

BIG SIDE EFFECTS

That societies, companies, and organizations have a need to measure and quantify to be able to function is understandable. The interesting question is, When does this no longer work? When do the numbers go from being performance enhancing to the opposite?

Several researchers have started looking at this, including where it concerns use of measuring systems and bonuses in companies. These research findings indicate that financial bonuses have a rather small and short-term effect and that bonuses can actually counteract their purpose. The results from the studies constitute a clear echo of Jordan Etkin's studies of self-quantification. Extrinsic motivation over time (in the form of bonuses) suppresses intrinsic motivation and may undermine the purpose—and effect—of the bonus.

It's not particularly difficult to point to several other—how shall we put it—"unintentional side effects" of measurement and quantification. This applies regardless of whether you measure and monitor yourself voluntarily or are quantified and measured by others. Etkin has shown in an elegant manner that even though self-quantification can cause a short-term boost in performance, measurement can quickly kill motivation and the will to continue performing. Another obvious side effect is that one becomes extremely ego fixated, in some cases to a point bordering on narcissism. Timothy Ferriss, from the start of this chapter, can have the honor of serving as an example of this.

A third unintentional side effect may be that you adapt your behavior to what can be measured. If your app can't count calories or the number of steps for a certain form of exercise, you simply omit that exercise. Otherwise the calculation will be wrong or incomplete. In companies and organizations this is a well-known problem, especially with regard to reward systems and key ratios. Employees gladly adapt and prioritize behavior in line with what is being measured and rewarded and put lower priority on other—often very important—tasks. As a related side effect, measurements can lead to cheating and self-deception. This can entail everything from shaking the phone with your hand in order to register more steps in the app to classifying ketchup as a vegetable when you count calories. Ketchup is mostly made of tomatoes, after all.

Another extremely common side effect of measurement is that you rely on the numbers even when you ought to ask yourself whether they might be incorrect or inexact. This can lead to what should be performance enhancing becoming the exact opposite. For example, if the numbers in your sleep app show that you slept poorly, you feel more tired and find yourself in a worse mood during the day—even if the app measured incorrectly and you actually slept like a log.

A final unintentional effect this measuring can have is that you become overly focused on improving the sort of thing you yourself have chosen to measure. If you check your weight and calorie intake, there is a risk both that you will diet too much and that your joy in life will disappear along with the calories.

So what number vaccine advice can we extract from all this?

1. Put away your measuring devices occasionally, unless you're an elite athlete or have medical reasons to use them.

2. Keep in mind that intrinsic motivation eats extrinsic motivation for breakfast. And carrots taste worse if you eat them to get thin.

3. Measurement can lead to decreased motivation and to self-deception. Be honest with yourself.

4. Keep in mind the story about Hugo Campos and his defibrillator. Your numbers are yours. Don't give them

to Fitbit, Google, or others without knowing what you get in return.

5. Keep in mind that numbers and measurements have strong unintentional side effects.

We're not quite finished with what you do and what the numbers do with what you do. Because they influence not only motivation and performance. They influence the experience of what you experience and learn about too.

NUMBERS AND EXPERIENCES

A few years ago I was the closing speaker at a big IT conference in Miami Beach. It took place at a fancy hotel that I never could have afforded to stay at if I had to pay myself, but since I was there for the conference, my family got a free stay for the week before, and it felt like I was floating on clouds. The hotel had everything: history, famous guests, a great atmosphere, and amazing surroundings! I remember how much I looked forward to telling my relatives and friends back home.

But when I'd checked out and was waiting for the ride to the airport, there was a ping on my phone: "Please rate your stay with us." I was being asked to grade everything from the room to the food, the service, the cleanliness of the accommodation, the sound level, and the environment, all with a number from 1 to 10. Even if the stay had been completely amazing, I couldn't give cleanliness more than a 7 because there were big leaves here and there that had blown down from the palms by the pool in the sea breeze; the same applied to the sound level, which was extremely high when performers were playing in the evenings. When I'd graded everything I ended up with a median rating of 8 out of 10.

And so that was that. I had just summarized this amazing stay as an 8, and suddenly it no longer felt as amazing. When I came home and people asked me what the conference had been like, I answered that it was good. "I stayed at a hotel that was an 8." No exalted discourse about everything I'd experienced, because an 8 isn't really all that, is it?

Rating literally diminished my experience. From having been something I looked forward to telling people about, something I struggled to describe in all its glory (presumably there would have been a bit of hand waving too), it was reduced to a single, sad number.

Micael

Why did this happen?

Because this is what numbers do: They summarize and reduce. They turn everything that's nuanced and rich into something simple and exact. Experiences are varied, but numbers are precise (they even have their own neurons).

Our experiences are built on a lot of different impressions, and they involve multiple senses: we feel, hear, see, smell, and taste. The combination of all these impressions makes experiences unique. That's what's so amazing about them. But this also means that they can be hard to interpret and explain, even to ourselves. And our experiences can be influenced by every conceivable thing. Not least what we *think* we ought to experience. Just take something like pain. It's not exactly a pleasant experience, but it's an experience nonetheless. Can we feel pain simply because we *believe* we're in pain? There is an interesting case, now 25 years old, in which a construction worker fell and landed with his foot directly onto a 15-centimeter-long nail that was sticking straight up out of a board. The nail penetrated right through the construction worker's boot and made him scream in pain. He was in such pain that the doctors couldn't see any other alternative than to get him the strong (and dangerous) pain-relieving drug fentanyl, which is up to 100 times stronger than morphine. But when they finally succeeded in taking off the construction worker's boot, it turned out that the nail had squeezed between two toes and that he was basically unharmed. The case was so unusual that it was written up in a scientific article in the *British Medical Journal*.

It can work in the opposite way too: we can feel considerably less pain than we ought to feel if we simply assume that everything is in order. Our experiences are not only highly individual but are colored by what's going on around us, how we feel, what we believe and see, and every other conceivable circumstance. It becomes almost impossible to compare with any precision one person's experience with another's.

This is also one of the reasons that patients get to report their own subjective pain when they're being categorized for treatment. The subjective pain can be reported in words (verbal scoring) or with numbers. And here it gets interesting: several studies compare the two ways to categorize pain and all draw two common conclusions. First, the two ways of scoring do not match particularly well; person A, for example, may use a stronger word for pain than person B, while person B sets a higher number than person A. Second, the variation is greater when patients grade with words as opposed to numbers. Responses based on verbal description spread out over several different word categories, from the weakest to the strongest words, while the majority of number-based responses cluster around a few numbers closer to the middle of the scale.

THE GRADED LIFE

Numbers therefore do exactly the same thing to pain as they did to Micael's hotel stay: they reduce the experience. Clearly, it can be a good thing to reduce the experience in the particular case of pain, but the point is that numbers even influence our medical experiences and conditions.

And, even worse, at least if you like watching movies, they reduce our film experiences too. A film manages to offer a lot—laughter, tension, surprise, and maybe even tears—for an hour or two, but when we rate the film afterward, all these impressions are reduced to one little number between (most often) 1 and 5. And the unpleasant thing is that the more often we put a number on a film, the lower the numbers become with time. American researchers found that pattern when they analyzed hundreds of thousands of film reviews on Netflix. Every time a viewer graded a new film, the probability went down a little bit that the viewer would choose a high number. As with the pain numbers, they started to gather closer to the middle of the scale.

And, even worse, the numbers reduce our happy experiences too. Micael discovered this when he asked 1,000 individuals to grade their perceived happiness in various domains of life, such as work, free time, health, and relationships, for several weeks. As the weeks passed, the participants set lower numbers on average for their perceived happiness in all domains of life.

Numbers take away that which is rich and unique in all experiences and make us think that they are exact and comparable. The more experiences we compare (and we do that every time we grade them, often completely unconsciously), the harder it becomes for each individual experience to stick out as something special and get a high number. In that way, our reference point is shifted so that a pleasant experience, which a year ago would maybe get a 4, now only counts as a 3. And because numbers are precise, and the number 3 is clearly lower than the number 4, pleasant experiences can end up no longer feeling pleasant at all.

Numbers transform us from curious participants in our own experiences to professional opinionators who always have the truth in the form of exact numbers with which to compare everything. And these professional opinionators get ever greater power over our lives in pace with the numbers sneaking into more and more experiences as we're asked to grade everything from hotels and movies to restaurants, doctor's trips, lectures (as scarred lecturers, we have cause to come back to that particular trauma), and even visits to the toilet.

And it's not enough that our inner professional opinionators take control of our own experiences; they take power over other people's too. Because the grades we give to the hotels, movies, restrooms, and all the rest often become numbers that are baked into the median value visible to others: "This hotel has received a 3.7 rating from other guests." Is the effect truly the same as when we assign grades ourselves? These are, after all, numbers based on others' experiences, not our own.

Unfortunately the answer is yes. We checked this out by having several hundred people taste a new chocolate bar.

One of Sweden's largest chocolate manufacturers was just about to launch a new flavor, and we got the opportunity to let people try it for the first time. In our study, half of the test participants learned before they tasted the chocolate that others had given it a rather low average grade, below 5 out of 10. The rest of the participants were told that others had given the taste a rather high average grade, way above 5. After that they themselves got to taste and grade the chocolate. The first group gave the taste a considerably lower grade than the second group. We also asked them to describe the experience in words, and those who gave the lower grade used rather lukewarm words, such as "okay" and "nothing special." The second group was considerably more inclined to use phrases like "really good" and "great." Everyone tried the same chocolate bar, but the numbers they got to see ahead of their tasting gave them completely different experiences.

Does this work the other way too? Can other people's numbers influence our experiences even in retrospect, when we already have made up our own minds?

This too we tested with chocolate: 100 people tasted the new chocolate bar, and once they'd done that, one half of the group was told that others had given an average number below 5, and the other half heard that the average number was above 5. The result was the same as in the previous test. Those who saw the lower number set a lower number themselves and used more lukewarm words to describe their experience, and those who saw the higher number described their own experience both with higher numbers and more enthusiastic words.

We think they're so precise, those numbers, that we're even prepared to revise our impressions of our own experiences after the fact.

DID I LIKE THIS?

Can this explain why we are so sensitive about how many likes we get on the pictures from parties, trips, dinners, and so on that we post on Instagram? Does the number below the picture speak to us about how good the experience really was? Most of us have probably at some point felt disappointment at having received considerably fewer likes than we'd hoped for after posting a picture of what we thought was a completely amazing experience.

> I'd been at a totally fantastic concert. It was a band I'd recently discovered, and I was surprised at how incredibly good they were live and by the powerful atmosphere they brought to the whole audience at a packed arena (we Swedes are not exactly known for huge public expressions). When I came home later in the evening, I googled for reviews from the concert. That in itself was rather strange behavior, since I'd been there myself and knew just how good it was. But I probably wanted to indulge a

> little more in the magnificence of the evening and extend the experience by reading about it. I clicked on the first hit on the Google search, which was a review in one of the major tabloids. To my surprise the review began by giving the concert the grade 3 out of 5. And even though I didn't agree with the grouch's description of the performance, choice of songs, or basically anything else, I couldn't shake off that number: 3 out of 5. I caught myself thinking that maybe the concert wasn't so amazing after all.
>
> Micael

Back to Instagram. We asked almost 2,000 people to describe their most recent post and grade the experience behind the picture in numbers and words. Half got to start by checking how many likes the post had received, and the other half got to check the number of likes only afterward. Guess what we found. The grade that the participants put on their own experience fit with how many likes the post had received: the more likes, the higher the grade. We guess that this is due to the fact that in the back of their minds, the majority remembered if the post had few or many likes; the majority of hearts come within a few hours after the post is made, so the memory of the post and the likes flow together. It's simply not really possible to disconnect the experience from the number in your head. The really striking evidence is that the connection became even stronger when the participants *first* got to check the number of likes. Then many likes led to even higher grades, and the participants described their own experiences in even stronger, more positive terms.

So much for "you had to be there."

The strange and creepy thing with how the number of likes influences our experiences is that the number has nothing at all to do with what we experienced. You and a grouchy reviewer were at least there at the same concert; the anonymous people behind the average ratings on the restaurant you visited had also eaten the same food or, in any event, been at the same venue. But the likes you get on your Instagram post come from people who weren't there, who didn't experience what you did, who don't have any idea what it was really like. Even so that number becomes a kind of template for your experience.

And it gets even creepier, because the risk is that you will let that number influence *what* you choose to experience next time and *how* you choose to experience it. Go back to your own Instagram grid and check. In the beginning you probably posted pictures of all kinds of events and experiences, some of which will have gained more likes. It's possible, maybe even probable, that a pattern will emerge in which the kinds of pictures you post more often are reminiscent of those you once received a greater number of likes for. So, in that way, the number of likes determines which experiences are worth telling others about—or maybe even worth experiencing again at all.

Likewise, have you ever caught yourself at a restaurant choosing an entrée based on how many likes you think it would get if you posted it on Instagram instead of how delicious it sounds? Maybe there are more people than we think acting in this way, choosing entrées based on potential likes from anonymous people who aren't even present rather than on what the server suggests (after all, they don't know what you like, and they've probably been told to recommend that, right?) or even on what someone in your own group suggests ("We don't have the same taste at all").

The number of likes of course takes away all the untidy and subjective stuff that makes our experiences unique. When someone else describes their experience in words, we can rather easily consider that as *just that* person's experience, *an* experience. But when the person instead uses a number, we suddenly see that as *the* truth, as *the* experience.

It's so bad that even the author of a book about how numbers are treacherous frets about having given a lecture some years ago that was sent all over the world and felt like a 10—but that, upon evaluation by the organizer, turned out "only" to have gotten an average grade of 7 from the participants. He checked the comments from those who gave the lower numbers that pulled down the average, and almost all were about the fact that their video and audio were out of sync. In other words, the low numbers didn't seem to have anything at all to do with his effort as lecturer. But why does that matter when they dragged down the average grade? Shaking off the comments about audio and video was easy. But the number is still there. He (who prefers to be anonymous) is actually still to this day a little ashamed and worried that people will see the grade and think that he's not a "top lecturer" after all but a fraud (and he feels like one, even now, when he's writing about the number).

If we let other people's numbers be the template even for our own experiences, how can we possibly not be easily influenced by the numbers relating to experiences we ourselves haven't had yet—for example, when we're choosing a movie or restaurant?

Think about it. It's probably happened before that you've been eager to see a movie but decided not to after you've seen that a grumpy reviewer gave it a low grade. And perhaps you didn't read what the reviewer wrote, or else you read it and decided that they were a grouch (since, as we've established, we all get like that when we grade too much), but you still couldn't shake off that number. Or perhaps you've avoided going to a restaurant because it got too many low ratings? ("The waiter didn't know anything about French wines," one review says—you don't even drink French wines, but the number is still a number!)

What hotel should you choose? The one that in the review had a great room and delicious breakfast? Or the one with a room that was so-so and a breakfast that was okay? That choice is probably rather simple. But what if you found out that the first hotel got a grade of 3 and the other hotel a grade of 5? Then the answer is no longer a given.

We couldn't help ourselves. We had 1,000 people read a

hotel review. In a variation of the review the (fictional) person offered some rather lukewarm words about the hotel (including that the breakfast was okay and the room was average), but nonetheless gave the hotel the highest grade (5). In another variation, the reviewer wrote extremely positively (including that the breakfast was delicious and the room was great) but only gave the hotel a 3. Even though the written review contained significantly more information about what was good or not, people were on average a bit more inclined to stay at the hotel rated a 5, meaning they were very clearly influenced more by the number than the written praise.

The deterrent effect of low numbers has given rise to an unpleasant phenomenon in which people deliberately set low grades in order to sabotage. Small businesses such as restaurants, cafés, hotels, and boutiques especially risk being affected, because they don't have as many customers to grade them, and every new number therefore has a rather large effect on their average grade.

But big companies have also been subject to what we call "sabo-ratings," when people give low grades in order to sabotage. It was front-page news when Facebook closed down a group set up with the express purpose of gathering people to give Marvel's movie *Black Panther* the lowest rating on Rotten Tomatoes in order to discourage people from going to see the first Black superhero. Disney has been subject to several similar sabo-ratings of their films, CNN's mobile app got thousands of 1s as a grade within 24 hours after publication of a negative article about Donald Trump, and the luxury hotel Boca Raton Resort in Florida saw its average grade deep-dive within the course of several hours after a YouTube star encouraged their subscribers to sabotage.

Sabo-ratings also create additional problems by influencing all these number-based algorithms that determine how high restaurants, hotels, and shops end up in the results lists and rankings on Google, Yelp, Tripadvisor, and the like. The algorithms rely more on numbers than on words, just as we humans do, and sift away anything with low numbers.

"The numbers don't lie," they used to say. Well, of course they do. Think about that next time you're going to grade something or check what others have given as grades.

And be aware that the grade you give influences what grade others are going to give the same experience, and vice versa. Because we're inclined to think of numbers as determining reality, we're also inclined to ourselves assign numbers close to the average grade, regardless of what we think about the experience (if we have even had the experience at all!). American researchers who analyzed movie ratings on Metacritic, book ratings on Amazon, and restaurant ratings on Yelp found that if dissatisfied individuals (perhaps grumpy reviewers or sabo-raters) got there to grade something first, people continued to give lower ratings than if someone perfectly content had given the first rating. People seemed to take the first average grade as a template and more or less copied it; when the researchers compared the rating with what people wrote in the reviews, they found that the text seemed to have little connection to the number (and a completely different effect on the actual sales numbers).

Back to that fine hotel, the year after Micael's stay there. I decided to give the hotel a chance, both after Micael's (immediate and verbal) recommendation and the hotel's positive ranking on Tripadvisor and Booking.com. Average rating of 8.1. Location: 8.6! Comfort: 8.6! Happy and expectant, the whole family arrives in Miami Beach and is welcomed by palm trees, long beaches, and radiant sun. The first thing I check in the Uber (besides the driver's score) is the hotel's address on Booking.com. And then I discover that the hotel's rating has gone down. From 8.1 to 7.9! The dream hotel has evidently been transformed to a shitty hotel over one night. And as I desperately browse among the numbers, I find out what's dragging down the judgment: "Wi-Fi connection" in the pool area has gotten the grade of 6.7. "Value for money" at the outdoor serving area is 7.6.

And what do you think happens then? Exactly: I devote major parts of the stay to being disturbed by poor Wi-Fi and warm white wine in the pool area for $17 a glass plus tip, while my wife and children excitedly run around the amazing pool area with their virgin piña coladas and the world's biggest smiles, happily ignorant of the hotel's falling rating on Booking.com and Tripadvisor.

Helge

700,000—AND YOU ARE IN LINE

We could have stopped here. But a pandemic intervened as we were writing. A pandemic that filled the news with numbers every day. About the number infected with Covid-19 and its many different mutations and about the number of reported deaths. It made us wonder and worry. If the numbers influence our experiences, even pain-related and medical ones, as we determined at the start of this chapter, how do all these coronavirus numbers influence how people are feeling when the number epidemic and the pandemic meet?

In the winter of 2021, we asked over 2,000 Swedes how healthy they felt right then, how great they thought the risk of getting infected was, and how worried they were about getting the coronavirus. One-third of them got to answer the questions immediately, while another third were first told the current number of infections, and a final third were instead told the current number of reported deaths.

Those who answered the questions immediately assessed on average that the probability they themselves would be infected was approximately 30 percent (and, interestingly enough, that the probability of the average Swede being infected was over 40 percent). This was much higher than the 7 percent who actually had been infected during the past year. Presumably the numbers they had already been bombarded with by the news had given them a vague sense that the infection rate was much higher.

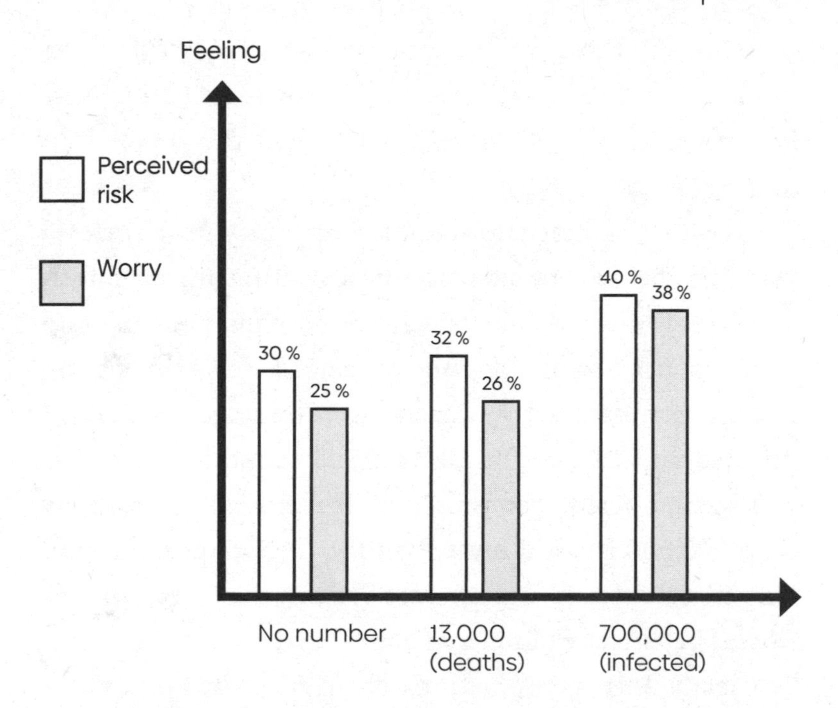

But the third who got to see the current number of infected Swedes (700,000) assessed the risk as approximately 10 percentage points higher, at 40 percent (both for themselves and others), and the anxiety they reported feeling rose to the same extent! In reality, the percentage of infected Swedes was actually barely 7 percent (as every Swede knows, the population of Sweden is just over 10 million). But as we determined earlier, we humans cannot defend ourselves against instinctive reactions to numbers—and 700,000 is of course a very large number, much, much bigger than nature intended that we should ever need to handle and understand. That also explains why those who instead saw the considerably lower number of 13,000 dead assessed the risk as lower and were less worried—but still more worried than the people who didn't see any numbers at all!

People's risk assessments and worry may have differed between the two numbers because getting infected feels more likely than dying (which we humans prefer not to think about). For that reason we redid the experiment and instead expressed the numbers as percentages: 7 percent infected and 0.2 percent deaths. The result? Putting it as 7 percent infected got people to feel less risk and worry than 700,000 but still more than the even lower number of 0.2 percent deceased. Both were *still* higher than for the people who didn't see any number at all.

Clearly, this says something about how hard it is for us to defend ourselves against numbers, when even a relatively small number is far more tangible and fear inducing than an unquantified and easily dismissed feeling that lots of people are affected.

This can explain why the number of people with symptoms of stress, depression, and poor mental health increased during the first year of the pandemic, why both actual and perceived isolation increased, and perhaps also partly why the authorities in Sweden, Norway, and other countries felt a constant need to hold press conferences and make policies. The numbers are too tangible and big not to react to them immediately.

And worryingly, these conclusions probably also indicate that the general numberfication of society, whereby we are constantly fed with numbers of every possible type, may have greater and more troubling effects on our well-being and our sense of security than we understood at the time. That unfortunately gives us reason to dig deeper into the pandemic in a later chapter.

For now, here comes a little number vaccine to remind you about the effects of numbers on your experiences:

1. Numbers reduce your experiences. Keep in mind that in the best case, they are an average of several dimensions and aspects of your experience (and in the worst case not even that).

2. Putting numbers on experiences does not make them comparable; all experiences are unique.

3. Be aware that both your own and others' numbers can rub off on your experience, both in advance and afterward.

4. Grading makes you fussy. The more number grades you assign, the lower the grades you give. For that reason be careful about assigning grades to everything and everyone.

5. Numbers contain not more information than words but *less*. Don't let the numbers replace other information; instead use other information to interpret the numbers.

We unfortunately have reason to add this bonus tip:

6. Numbers can not only influence how you experience pain but also even make a whole pandemic worse. You literally need to vaccinate yourself against them just as you get vaccinated against the virus.

If numbers influence experiences, which we often share with others, the natural question is whether they influence our relationships too. If the number epidemic is, in a way, just as contagious as a viral pandemic, do we infect one another with our numbers too?

NUMBERS AND RELATIONSHIPS

At the end of September 2015, Peeple became the internet's most hated app before it was even launched. The *Washington Post* published an article about the app, which had already been valued at almost $8 million, and described it as "Yelp for people." In the same way that Yelp lets people grade businesses, Peeple would make it possible to grade other people on a scale of 1 to 5, either professionally, socially, or romantically. "People do so much research when they're going to buy a car and make similar decisions. Why not do the same kind of research in other aspects of life?" the founders asked, then explained that the app would be perfect both for sharing your personality with the world and finding people who can be trusted. "We want to spread love and positivity."

But they didn't get much love and positivity in return from the journalist, who concluded the article by calling the app a "terrifying" and dystopian vision of the future. Or from the rest of the world's media either, in broadcasts and newspaper pages, or in the storm that soon broke out on social media, where the founders were more or less bombarded with hateful comments.

The commotion led to postponement of the app's launch, and when it was finally released six months later, it was different: people could choose if they wanted to be graded without numbers and if they wanted their score to be visible. It got rather tepid reviews, and Peeple has since led a relatively obscure existence.

In the end the *Washington Post*'s dystopian vision of a future where we grade one another by means of an app didn't become reality.

Instead the reality got much worse.

Because instead of an app where we grade one another in three defined ways, we now have hundreds of apps and "services" that put numbers on and influence our relationships in every way you can imagine. You can grade the clerk who just helped you at the shoe store. The doctor who just wrote you a prescription. Your yoga instructor. The coach of your soccer team. Or your teacher. On RateMyTeachers.com and

RateMyProfessors.com, students have assigned millions of numbers to their instructors.

And because teachers and professors are academics, of course they've taken a closer look at these grades and found that, to a large degree, the numbers seem to say more about the grader than about the instructor—whether, for example, a student was satisfied with their own grade in the course, had just that day been told off for turning up late, or (strikingly often) found the teacher attractive (which, by the way, was a separate category on RateMyProfessors.com until 2018, called "hot chili pepper").

To be continuously ranked and graded as a professor requires a strong stomach and a rather stable psyche, I would say. You're not simply being number-judged on RateMyProfessors.com and other websites, but at most universities and colleges every course is also internally judged by the students. Especially in small classes, a single bad numerical judgment from someone unhappy can completely ruin the whole average evaluation of a course. And when the motivation behind the grade concerns either the quality of jokes, accent, or appearance ("are you going bald?"), it's even more irritating. Believe me. It is the same with students seeking revenge for slightly stricter treatment from the lecturer. A charmer who "forgot" to fulfill the requirements to be able to take the exam once gave me the following ultimatum:

either let them take the exam anyway or receive a crushing 1 in the course evaluation. I answered quickly and resolutely and, hardly surprisingly, just as quickly got back a 1 in the course evaluation. A kind of quid pro quo, I guess.

Helge

And does the same logic apply when you grade your boss (there's a group of "services" for that)? Or your colleague? Your classmate? Your date?

Imagine if these grades influenced us in the same way with dating as with choosing hotels. Whom are you more inclined to swipe right on: a person whose dating profile you think is *quite* attractive or a person whose dating profile you think is *very* attractive? Just as with the hotel reviews in the last chapter, this choice is probably rather simple. But again, just as with the hotels, it gets trickier if we say that the first person, the one who wasn't so attractive, has a grade of 5 stars on their profile, while the other has a 2.

When we randomly assigned grades on 100 dating profiles so that they either got 2 stars or 5 stars, people's inclination to swipe on them also changed. When a dating profile got 2 stars, the number of left swipes ("no thanks") increased by between 25 and 30 percent, and when it got 5 stars, the number of right swipes ("yes please!") increased just as much—regardless of how attractive the people in the profiles were. When we placed more attractive individuals with lower scores against less attractive individuals with higher scores, participants *were* more inclined to choose attractive people, but the difference was not that great (unlike when they weren't graded)!

As we've said, we struggle to defend ourselves against numbers. Reflexively we shy away a little from those who've gotten low scores and are drawn to those with high scores. And we feel the numbers assigned to ourselves almost physically.

The tricky thing is that this part of the brain where the number neurons are located, the intraparietal sulcus (IPS) as you might recall, doesn't only process numbers and body movements. It also processes how we interpret other people's intentions, research shows. Why that's the case is not crystal clear (it seems that the IPS could be a catch-all for a little bit of everything), but it probably has something to do with the fact that it's crucial for our survival to be able to figure out what others intend to do—if they are friends or enemies, want to help us or injure us—in the same way that it's crucial to be able to quickly react to different quantities and sizes. As you recall, we program our brains to connect quantities and sizes to numbers that make us react faster than we're even able to think. The risk is that we do the same thing with our brains when we grade, so that we more or less automatically interpret our own and others' grades as signals about what we actually think and mean.

RATENAPPING

As if it's not bad enough that we can't really defend ourselves against the grades we assign to one another, there is also the risk that we're starting to look at one another in the same way we look at movies and other experiences—like critics who give stingier and stingier grades.

How does this actually influence our relationships and the way we conduct ourselves?

The first time my son and I took an Uber together, he asked as we were getting out if all the drivers were as nice as the one we'd just ridden with. "Much nicer than taxi drivers usually are," he said happily. He was a little disappointed when I took out my phone and explained that, yes, Uber drivers are normally nicer, but that's probably because we now assign grades to them, and they would surely like to get a 5.

"I see. But you were also nicer than you usually are," my son said, shrugging his shoulders. "Do you get a grade too?" I was on the verge of saying that it was probably the driver's niceness rubbing off on me, when I was struck by a thought and forced to look at the phone: yes, I had received a grade too.

Since then I always feel a certain degree of performance anxiety when I get into the back seat. Because it's not enough that I have to pay for the ride; I have to perform as a passenger too and be pleasant so that I don't get a low grade. Otherwise maybe no one will want to drive me next time.

Micael

Would you dare give anything other than a generous tip if you thought the driver could give you a low grade as a passenger? There are reports of passengers who, when they're about to get out of the car, hear that if they don't give a big enough tip, they're going to get a low grade. The same if they don't give the driver the highest grade. Even the perception that numbers could be used against them in that way changes the way they act. This grading behavior, consciously or unconsciously, seeps into our other relationships too.

When Snapchat wanted to get people to use the app more often, it launched "Snapchat streaks," which showed how many days in a row two people had sent snaps to each other. When the numbers reached a certain level, those users got a pretend trophy, but as soon as they missed a day, the number was immediately nullified. It quickly became a thing among young people to send black snaps to one another to keep the number steadily counting upward. Sure enough, these black squares increased the number of snaps that were sent, but they also meant the snaps didn't contain pictures or messages, simply becoming empty number values.

As a parent it's not always that easy to know about all these functions in the apps or how you should handle them. One

> evening in 2017 my oldest daughter became unusually upset when I nicely but firmly asked her to put away her phone and Snapchat at bedtime. I was told that I was ruining her life if she didn't get to fulfill her streaks. At that time I associated the word "streak" with individuals who ran naked in sports arenas, so I didn't understand either the relevance or her intense engagement with the streak. It turned out that she had devoted weeks and months to working up a long series of streaks with a ton of friends on Snapchat—streaks clearly with an enormous value that I was now going to entirely sabotage in a single evening. The world's worst dad.
>
> Helge

Soon the media started reporting on stress and anxiety among young people who manically followed their numbers. Some begged and pleaded with their parents to send blank snaps in their place when they themselves were occupied (for example, with very minor distractions such as going to school) or lacked Wi-Fi; others became hostile when a friend missed their turn. Some felt compelled to snapchat despite having no interest in the game, and some were worried about not having a streak partner with whom to hit the high numbers.

RELATIONSHIP, PERFORMANCE

But the numbers have also wormed their way into every conceivable app for adults. Like relationship apps that calculate how many times you and your partner send messages to each other or that encourage and count romantic gestures (in case you aren't aware of them already, we won't name them out of pure consideration for your relationship, but unfortunately they are only a Google search away). Or sex apps that help you log how often you have sex, for how long, and how good it was. All have the benign intention of increasing *quality* in your relationships but instead risk focusing your attention on *quantity*.

Because who doesn't understand that four romantic messages are twice as good as two or that eight minutes of sex is better than seven? Perhaps you even think that eight minutes sounds short when you see the number? But it still constitutes "overperformance" by almost three minutes according to research that found the average intercourse takes five minutes. Presumably you would also think having sex once a week was way too infrequent if that's what your counter showed, even though it would be "overachievement" compared with the average of 0.75 times per week in a British study. And you'd probably feel dissatisfied with once a week, even though research shows that's the optimum amount and that couples who have sex more often than that aren't any happier because of it.

The risk is that the numbers transform relationships into

performance. And, as we're sure you recall, numbers have a tendency to influence us to perform more and at the same time feel less happy, until, in the worst-case scenario, you're sending romantic messages to each other and having sex for the sake of the counting rather than because you really want to. Until the only difference between you and the young people on Snapchat is asking your parents to send romantic messages to your partners when you're occupied by a work meeting.

So do numbers transform relationships into performances? The question was too disturbing not to seek an answer to it. What happens, for example, when we get numerical help with our dating lives in the form of an average score on the people we're swiping?

We conducted an experiment to investigate this. We asked 1,000 people eager to date to test two different versions of a known dating app. Half got a version with average scores on all the profiles they viewed, while the other half got a version without average scores. Afterward we could see that those who got the version with scores looked at more profiles but still had shorter sessions—as if they were managing a job as efficiently as possible. When they responded to our survey afterward, sure enough they answered higher on the scale when we asked if they perceived the swiping as a job. They also thought that it wasn't as sexy or fun.

But back to those romantic gestures. The danger is that we will start competing with each other. That may sound like a pleasant thought—trying to outdo each other in love bombing—but if your partner sends you three romantic messages every day, chances are you will feel stressed or guilty about having "only" sent two. Or, worse, you'll feel pressured by that pest who's going to win every day. And it works the other way too: Could your partner start seeing you as the teammate who coasts in the relationship instead of contributing equally to the numbers?

Worst of all, the relationship could cease to exist at all.

The most common Tinder-related searches on Google are "how many swipes per day" and "how many likes per day." "How do I match with someone" and "How do I meet the right person" don't even show up among the top-10 most common Google searches (on the other hand there is "how many matches per day").

Studies on Tinder users show that many use the dating service as a pure ego boost or as entertainment, without wanting to actually meet someone; the goal instead is to get as many likes and matches as possible. That would explain why as many as 55 percent of those asked in an American study answered that they use Tinder *and* that they already have a partner (the number rose to over 70 percent in another study that instead asked people if they'd seen anyone else on Tinder who they knew wasn't single).

It would also explain why studies have found that certain people seem to be dependent on Tinder in the same way as others become dependent on poker. The heavy Tinder users talk about their "Elo," which, put simply, is a number for how many swipes they themselves get compared to how many swipes the people they choose or reject get (the number comes originally from the chess world as a measure of how much a win is worth depending on how many wins the opponent previously had). The focus on the number of likes and matches instead of actually meeting someone could also explain why research has found that Tinder use could lead to people becoming less satisfied with their appearance and having lower self-esteem.

Tinder is far from the only number generator that influences our relationships. Guess what one of the most common searches on Google is for Instagram: "How do I get more followers?"

The app started as a way to share snapshots from life with friends and acquaintances, the modern version of the photo albums of yore that people looked at together with their near and dear. But over time it has become a bit of a follower accumulator. For many people the number of followers and the will to increase it are hard not to focus on. It's gone so far that there is a whole cluster of various services where you can even buy followers (the same thing can be arranged for those of you who prefer Twitter).

Surely you keep track of how many friends you have on Facebook too? And how many contacts you have on LinkedIn? Most people do; we asked. We requested 1,000 randomly selected individuals to tell us how many friends they had on their social media. All responded with exact numbers. (And yes, we can give you the averages, because of course you can't resist the impulse to compare them with your own: Instagram 167, Facebook 755, Snapchat 47, LinkedIn 353). In addition, they seemed to think that it was easy to determine how many social media friends they had; we asked about that too, and on a 7-point scale (where 1 is very difficult and 7 is very easy), the majority rated the task a 6.

If instead you were asked "How many friends do you have?" or "How many business acquaintances do you have?" in the "real" world rather than on social media, you would presumably have a considerably harder time answering. People do. The same 1,000 people were more inclined to answer with a rough guess instead of exact numbers (the average ended up as the round number 20 for friends and 50 for business acquaintances), and they thought that it was considerably harder to determine how many friends they had—on the 7-point scale comparing them with others, the majority only put a 4. Because why would they keep track of how many friends and business acquaintances they have—think about it, what significance does that actually have?

But when the numbers are suddenly there, they become significant. They make the number of relationships important. And the numbers make the relationships exchangeable, because numbers are, of course, *just* numbers, to the extent that some are prepared to buy followers. They also make relationships comparable. Who has the most friends on Facebook, the most contacts on LinkedIn, the most followers on Instagram? Suddenly we're arriving at such twisted conclusions as we aren't good enough people if we "only" have 2,000 (!) friends (?) because we see that someone else has 5,000.

We risk becoming competitors in our relationships.

> I have an extremely ambivalent relationship to new social media. While I think it's cool when there's something new to try, I feel a little stressed. Do I have to start over from zero again? When I got an invitation to be part of the conversation platform Clubhouse, which was then completely new and extremely hyped, I found myself hesitating. I checked who was there, many "hip" names with lots of followers already. How would I ever catch up with them? How bad would it look if people checked up on me too and saw that I had almost no followers at all? Instead of being happy about the exciting new functions and possibility of talking with and listening to people from all over the world, I was worried about my pitifully low new number.
>
> Micael

Perhaps it is more than a coincidence (or, in any case, an extremely interesting one) that the number of single-person households has increased at almost the same landslide pace as numbers have wormed their way into our relationships. In

Sweden the number was 12 percent around 1950; according to EU statistics, by 2017 it had risen to over 50 percent! That makes the Swedes the world's most single people, but Norway isn't far behind at just over 40 percent, and the entire European Union has seen a similar increase over time, averaging just over 30 percent.

There are many factors behind this rise, but what if the high numbers on our dating, colleague, and friend accounts are getting in the way of our relationships in the same way as economic research has shown that our bank accounts do?

We can't help wondering if the numbers reduce our confidence in the people around us. If we assign numbers to

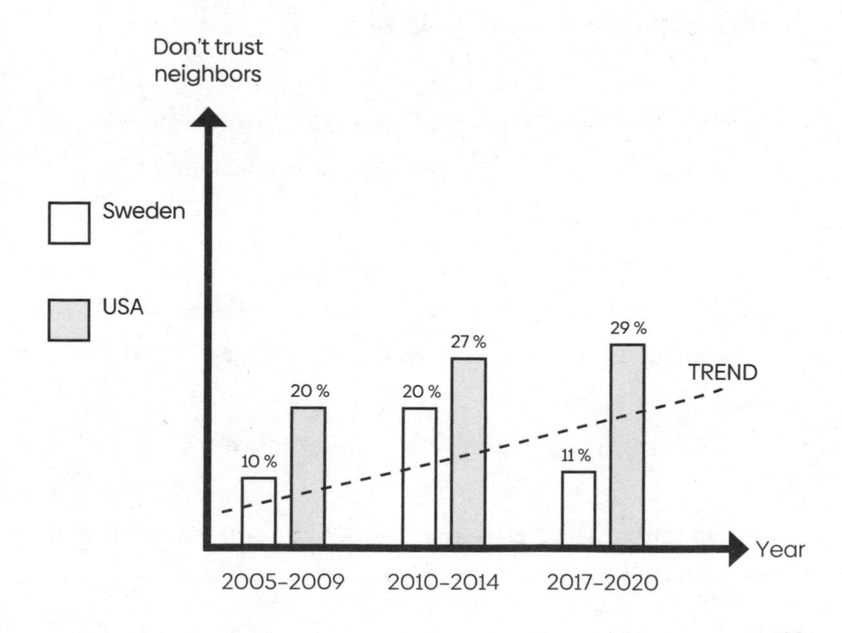

and grade our fellow humans in various ways and make one another counterparts in different transactions, isn't the risk probably rather high that we're starting to rely less on one another? What if the numbers reduce our empathy with one another?

Here's your number vaccine for relationships:

1. Differentiate between number and intention; they are not the same thing. The grade you get (and assign yourself) doesn't need to mean that the person really thinks that.

2. Differentiate too between number and quality. Your friends are not less worthy because they are fewer in number!

3. Keep in mind that your relationships are not performances simply because you now have the numbers on them.

4. Be aware that you can be kidnapped or kidnap others with number ratings, whether or not that's the intention.

5. And please, please, don't grade your professors.

If you truly can't help caring about the numbers, bear in

mind that six minutes is not short but long and that once a week is quite enough.

But we can't stop here. If numbers transform our relationships into performances and transactions, does that mean numbers have become a kind of currency in themselves? Let's look more closely at that.

NUMBERS AS CURRENCY

In 2018, one of North America's largest life insurance providers, John Hancock Insurance, announced that from that point forward, they would only be selling "interactive" life insurance policies that collect health data through wearable activity trackers. The customers, by giving the insurance company access to their health data via Apple Watch or Fitbit, would receive discounts and various benefits. Or, put differently, the customers were potentially punished, risking a higher insurance premium, if they didn't. At about the same time, Australian life insurance companies launched corresponding "innovations" giving benefits to customers who used activity meters and a bonus if they had a BMI below 28.

Motivating and healthy, the insurance companies thought. Dystopian, perverse, and invasive, critics believed.

All numbers about ourselves, whether we register them on our own or let others do it for us, have bit by bit gained great value—value for us, for employers, for the government, and not least for commercial companies. By gaining access to geolocation data, health data, numbers of likes and followers, and sensors in the home, in the car, and on our bodies, technology companies can give us better advice, more personalized services, more accurate advertisements, better risk management, and cheaper insurance.

The algorithms are self-improving via artificial intelligence (AI) and so-called deep learning. The latter of these, deep learning, is basically a neural network that allows technology to learn from large amounts of data in a way that mimics the human brain. The tricky thing with predictions via deep learning is that we humans have no insight into what data or rules the AI actually uses in its predictions. Even more worryingly, the companies that use the deep-learning models don't either. For that reason, the models are often called black boxes. And it's not ethically unproblematic if, for instance, your insurance premium is determined by a function of ethnicity, exercise patterns, and weight.

A Nordic bank was recently forced to retire its new, snazzy, deep-learning credit model, which predicted when people were likely to default on outstanding loans better than any other model or method. Why was it forced to do that? Well, because the bank neither knew nor could explain to the Financial Supervisory Authority what decisive criteria the model used as a basis for refusing someone a loan. And so here we are again: *computer says no*. Period.

We said in the foreword that we are in the process of becoming number capitalists where number of likes and followers, heartbeats and steps, bonus points and restaurant rankings are concerned. And we mean the word "currency" here in both a literal and a metaphorical sense. Number of likes is money, and number of followers is your bank account. If you're a blogger or influencer, followers and likes can literally be counted in dollars and cents. And pulse, steps, and elevation gains can be converted into an insurance discount. But numbers are also currency in a metaphorical sense. They are status, self-confidence, and negotiating strength. They can also corrupt the same way money can—in exactly the same way, actually.

If decades of research on the psychological effects of money have taught us anything, it's that money guides people's thinking and behavior. Just looking at or touching a bill can, as we've said, make people more egoistical, self-centered, and cold. We called this the "asshole effect," remember? When exposed to money, people adopt a more transactional way of thinking, become less inclined to help others, and make more egotistical choices. More recent research has also shown that people who have been exposed to money cheat more, share less, and make poorer moral choices.

Is it possible that numbers—as currency—are doing the exact same thing?

THE NUMBER NARCISSISTS' MORAL COMPASS

To investigate this, we sent out questionnaires to 800 Norwegians. First, we asked if they kept track of numbers concerning themselves: Did they log their own health data? Did they know how many friends and followers they had on social media? And did they follow financial data about themselves—that is, did they keep an eye on stocks, funds, bonus programs, and balances? After that, we gave them various moral dilemmas and measured to what extent they were inclined to take "shortcuts" here and there. These dilemmas were about everything from filching a little copy paper at work to bumping someone's car or getting the wrong (i.e., too much) change back after standing in line a long time to buy coffee. These are amusing and proven tests for moral choices. And what do you think we found?

There was a weak negative connection between registering your own *health data* and morals, so frequent Fitbit and Strava users are a tad less moral than everyone else. They are also a bit more self-absorbed than those who don't follow any activity meter.

The results for those who monitored their own numbers

and likes on *social media* were even more depressing. They not only reported higher stress levels but also came off considerably worse where moral dilemmas were concerned. They thought that it was okay to steal a little from your employer, make pirate copies of software, and keep too much change.

We also found the same pattern for people who monitored their own *financial* data. They came off worse in the question of moral dilemmas, prioritized work ahead of spending time with family and friends, and were even more likely to be xenophobic. A tasty cocktail, isn't it?

The more you keep track of your... **the**	**social media numbers**	**health numbers**	**economic numbers**
less independent and capable in life you feel	✓		
higher stress level you have	✓		
more unethical choices you make	✓	✓	✓
happier you feel		✓	
more social you plan to be tomorrow		✓	
more you plan to work tomorrow			✓
more skeptical you are about immigrants			✓

American researchers often explain the negative effects of money on people's morals by what they call a "self-sufficient mindset"—that is, when you have a lot of money, you are more independent and feel you get by just fine without help from others. Does that ring a bell here too?

That was exactly what we found in the studies where we fiddled with the numbers concerning how fast people had run (faster or slower than the average). If we let them believe that they had performed better than they really had, their self-confidence shot to the top, including a feeling of "self-sufficiency." They also got better results where risk-taking was concerned.

And when we let them respond to the same moral dilemmas, what do you think happened? Well, because of the high numbers, they felt superior and stronger—and were generally more open to acting a bit unethically in different situations. Exactly like people who are exposed to money. And exactly like people who have found out that they have lots of likes on an Instagram post.

So it's not just money that disrupts our moral compass; other types of numbers do that too. The number doesn't even need to be about anything; it can simply be a number or a math problem. In a series of experiments, researchers from Hong Kong and the United States found that people consistently become more egoistical, dishonest, and self-centered if presented with a math problem. The experiment was simple. The participants were randomly divided into two groups. One group had to solve a verbal problem, the other a numerical problem. After that they got to play a game, a so-called dictator game, where they had the option to keep more money than other players and to lie. The participants who had to solve the number problem consistently lied more and kept more money for themselves. Sad, but true.

These experiments also clearly point to the fact that we humans treat numbers and words differently. Numbers are currency that make us more focused on ourselves, more impersonal, and less emotionally engaged, and they bump our moral compass in the wrong direction.

During a period when I led a lot of long meetings, I amused myself by systematically observing whether the meeting participants poured coffee for one another or not. At meetings where there's a thermos of coffee on the table, you can choose to pour only for yourself or to ask if others around you also want some. And you can probably guess what I noticed. Yes, when we discussed numbers, budgets, and rankings, people mainly poured for themselves. More qualitative topics and documents entailed more social pouring of coffee and acknowledgment of others. Some even sent the candy bag or cookie plate around now and then. I also noticed, by the way, that some professors, who shall remain completely anonymous, who, I knew already, were extremely interested in impact factors and h-index and how often they were cited, and who frequently sharpened their numbers on Google Scholar and ResearchGate, were the first to slip out when new tasks or chores were assigned. I'm not saying that this was a scientific investigation but the observations were pretty cool as anecdotal evidence goes.

Helge

GAME ON!

"The most important thing in my game designer toolbox is the point system because it tells the players what to care about." These are the words of Reiner Knizia, a world-leading game designer, who has designed games such as *Lord of the Rings*, *Keltis*, and *Lost Cities*. Point systems in games seem to change people's mindset so they escape reality and become more motivated, narrow-minded, competitive, and, once in a while, so immensely frustrated that they toss tables and yell at one another. Imaginary numbers and points, without any value in real life, may make even the calmest and most reserved person suddenly fierce. And according to game philosopher C. Thi Nguyen, the logic of point systems in the gaming world is now eagerly being adopted and applied in society at large. Corporations, institutions, and even schools have started to understand how games and point systems can be used to shape our agency and behavior. We "gamify" everything from schoolwork, tax reporting, sales contests, and bonus programs to Twitter conversations. And as Nguyen puts it, "We don't play the game. The game plays us."

Numbers and point systems transform physical and social phenomena into units that are measurable. Your financial responsibility is converted into credit points, your social network into number of followers and views on social media, your wanderlust into frequent flyer miles, and your joy in exercise into calorie consumption and average walking speed per kilometer.

Numbers thereby increase competition and rivalry. By quantifying our entire lives, we now introduce competition into more and more arenas. Formerly qualitative differences between people and experiences that could be interpreted in numerous ways are now transformed into rock-solid quantitative difference. Two selfies, beach bodies, or dinners can suddenly compete and be mercilessly compared.

The same thing applies in the business world where a shopping experience is reduced to 3 stars, a restroom visit to a smiley, and a book or concert to a rating between 1 and 6. Numbers and quantification reshape complex phenomena into one-dimensional scales where much of the content is lost upon measurement.

Numbers therefore also influence language and the value of experience. "How good-looking was she on a scale from 1 to 10?" By connecting a numerical result to a quality, thing, or person, we also give an explicit assessment of value. An 8 is better than a 7. Quantifying makes the value simpler to relate to, things become simpler to compare, and we are all clearly ranked. Micael has 28,400 followers on Instagram; Helge has 135. Quantification makes social status clearer, and it's easier to convert social phenomena into hard currency. As where money is concerned, a higher number is normally better—perhaps with the exception of pulse and blood pressure. Algorithms digest big data and present them to you because you simply can't keep from clicking and comparing. You are 334, Nils is 176, your neighbor 189, and your partner 544. The number can apply to anything from social intelligence to attractiveness, ranking on social media, fatness, or depressive tendencies.

With new ways of connecting numbers, new services are created where numbers' role as currency becomes even clearer—and more absurd, some would probably say. For example, on the service CreditScoreDating.com, launched in 2006, you can find your future spouse based on your credit score compatibility. As it says on the website, 57 percent of all men and 75 percent of all women put emphasis on financial security when they choose whom they'll go on a date with, so what better way could there be to find your perfect partner? Birds of a feather...

Did you know, by the way, that in 2015 Facebook had already patented a method for calculating users' creditworthiness based on their social network? The underlying logic of the method is that if you have a cache of friends with little willingness to pay and similar low capacity, well, then there's a great probability that you have a poor credit rating too. So be careful whom you socialize with, both in person and digitally. Otherwise the numbers might be coming to get you.

MONEY IN THE MATTRESS

Back to the Quantified Self movement. The self-tracking advocated by Timothy Ferriss and others is blamed by many for converting numbers to currency and competition and thereby making every person into a little company. If all numbers from activity meters and smartphones are going to be used to optimize and improve their own performance, you're dangerously close to a market analysis of us people too. There is always a number to strive for, always something to compare and improve. Market logic gains the advantage over relationship logic, and people become small self-optimizing companies.

And because all these data about yourself also have great commercial value, exchanging them for new services and better advice from technology companies like Google, Strava, Facebook, and Apple is easy. If Google could access data from sensors on your body, from your smartphone, and from sensors in your home, your everyday life could look very different. Then the coffee machine could be turned on the moment you wake up, the house and the car adapted and programmed according to the day's tasks, your exercise program and mealtimes customized, and your most valuable relationships maintained and optimized. The more data you're willing to turn over, the better and more customized the advice. You might also receive more benefits in exchange.

For example, there are services for DNA profiling online. Yep, for about $100 you can get your own DNA profile mapped on the internet. And even better, you can upload it to various technology companies and get tailor-made advice about basically everything: dieting, exercise, baldness, pimples, gambling, freckles, aggression, depression, sunbathing, and coffee intake—just to name a few. There are also apps that, based on your DNA profile, can find out what wine would best suit your taste. Clever, huh?

And the struggle over access to your numbers and your data means that the technology companies are now moving into completely new industries that you might not have thought would be of any interest to them, such as beds and mattresses. Yes, tech investors have started shoveling large sums into something as analog and boring as mattress manufacturers. Could you have imagined that industry would be "disrupted"? And why mattresses?

Well, the investors imagine that in the future you won't buy a bed; you'll buy sleep quality instead. People don't need beds; they need sleep. Through sensors and monitoring of the mattress, your sleep can be optimized. You can even take that good sleep with you when you check into a hotel, stay at an Airbnb, or sleep in a tent.

NUMBER CAPITALISM

It used to be said that time is money. Now apparently numbers are money. And money saved is money earned. Bonus points can be exchanged for vacations, flights, or goods. Fitbit data can be exchanged for lower premiums on life insurance. A higher satisfied-customer index and other optimal measurements at work can earn you a bonus. Numbers about driving patterns from your car can get you cheaper car insurance. Better creditworthiness gives you better interest rates on loans. High ranking from restaurant guests leads to increased turnover. High social points in China give citizens faster internet. And the number of likes can be converted to shiny dollars and bitcoins.

Test this by googling "TikTok money calculator." (That "google" has become an accepted verb is incidentally in itself a sign of the power of number capitalism.) You'll get over 10 million hits. There is an endless series of calculation programs for converting the number of followers, views, and likes to quick cash: *How much money can you make on TikTok?* Admittedly, you need tens of thousands of views before you earn enough for a chocolate bar, but still—the currency is there. And millions of young people around the world dream of a future as an influencer or celebrity where likes and money flow in steady streams.

In 2019, 19-year-old Addison Rae Easterling earned more than any other TikTok star. Sixty million followers saw to it that $5 million rolled into her account. Does that sound wild?

Number 2 on the list, Charli D'Amelio, was even younger. At age 15, she had 8.6 billion likes and $4 million in revenues. Today she has well over 100 million followers. Her "career" has catapulted at record speed from TikTok child star via *The Tonight Show Starring Jimmy Fallon* to contracts with Prada, Hollister, and the Super Bowl. Her big sister Dixie is number 3 on the list, by the way, with 49 million followers and 8 billion likes. Is it really that strange that today's kids are so fixated on numbers?

And that doesn't just apply to kids, as we know. We are number animals after all, and the numbers automatically get us excited. They are God and mammon and porn at one and the same time. Their currency is built into our bodies, our brains, and our common history.

> Recently I posted an exercise video on Instagram. Nothing special about that; I do it fairly often. But the unique thing about that video was that within the course of a few days it got over 30,000 views, far more than the 10,000 the videos usually get. Soon it was upward of 40,000. In hindsight, I realize it must have been down to timing, since it was right at the start of the vacation period when people have nothing else to do than mindlessly scroll Instagram and, besides, were more eager for exercise inspiration. But the next time I made a video, exerting myself a little more and hoping I might reach 50,000 views, it was a disappointment: the same old 10,000, which no longer felt cool at all.

And I caught myself getting the same feeling every time I posted a new video after that; the response to that one video had made me greedy. Even if the views sometimes doubled, it still felt pretty puny compared with the 40,000 I'd once gotten. So much so that I stopped making more videos after a while, feeling like, what's the point?

Micael

If you think that money and the hunt for more of it can make us intoxicated and dependent, then you should probably also ask yourself what all the *other* numbers do to us. Ask Torbjørn Høstmark Borge how exhilarated by and dependent on his Strava numbers he became before things went out of control and his legs collapsed totally. Ask Parvez Iqbal about how exhilarated by and consumed with getting likes on his TikTok videos his son Noor became before he took his own life.

You see, the numbers are sneaking in as currency in far more areas of life than money. Nowadays you can take any letter in the alphabet whatsoever, and you're almost sure to find a counting device or service that starts with it. Do you choose *T*? There's Twitter, Tinder, TikTok, and Tripadvisor. What about *B*? There we have BMI, Betsson, and Booking.com.

Shall we continue?

Just admit it: you've become a number capitalist who wants more, higher, and better all the time. The numbers you strive for can be exchanged for everything from social status and self-confidence to customized services and financial benefits. The numbers make you both exhilarated and excited but, unfortunately, also a tad less moral and social. Oh well, since the numbers have become so important to you, it's good they are so concrete, objective, honest, and true... More about that in the next chapter.

My son Dante and I went to the movies when they finally opened up the movie theaters for limited showings again during the pandemic. We both love going to the movies and were excited to see the movie *Tenet*, which had quite a delayed release date. We both liked it a lot. I chose not to check the reviews afterward (I've learned my lesson), but I couldn't keep from clicking on a headline saying that ticket sales for the movie were a disappointment compared with expectations. The disappointment resulted because the film was not anywhere close to breaking a record for ticket sales but instead just barely made it onto the list of the 10 most revenue-generating films in the past 10 years. I thought it was bizarre, first, that a film's not breaking records was a disappointment (not every film can break records, right?), and, second, that a film that nonetheless sold so many tickets *in the midst of a pandemic*, when cinemas were at less than half capacity, was considered a disappointment—how number greedy is that? But the most bizarre thing was probably that I found myself also being disappointed that the film we liked so much didn't get higher sales numbers. "Wasn't it better than that?"

Micael

Well then, now we've determined that we've all become number capitalists. But is that perhaps a bit too gloomy and dystopian? A future world where numbers from your Fitbit, your phone, your mattress, your social media, your car, and your home are transformed into discounts, money, status, and immorality? How do you feel about a little pick-me-up here at the end of the chapter? Don't numbers acting as currency also have positive effects?

Of course they do. Because we often believe more in numbers than in ourselves, numbers can also save us from situations where uncertainty and prejudices bring out our worst.

It's well known that innate prejudices play a role in how we think and act with respect to people who are different from us. We know, for example, that Airbnb hosts who belong to a different ethnic group ("them") are paid less than hosts who are "like us" (the same ethnic group). We checked this out in Norway. In three experiments with a total of 1,600 participants, we tested how Norwegians reacted to completely identical apartments but with different hosts. The results were disheartening. When the same private homes were presented with a representative of a non-Western minority as landlord, the participants were more negative about the apartment, and the probability that they would choose it was as much as 25 percent lower.

What happens if we then introduce numbers in the form of a rating, 1 to 5 stars, based on other guests' experience? Does that help? Yes, *a lot* actually: with a 5-star rating, all uncertainty and prejudice disappeared like dew in the morning sun, and the difference of 25 percent in the choice between ethnic group and group host sank to zero.

So numbers as currency not *only* have dark and dystopian sides, thank goodness. They also guide us and increase our feeling of control in situations where we risk otherwise making decisions based on prejudice and uncertainty.

Even disregarding the positives, number capitalism is not easy to vaccinate against. You can't just turn off the world, move out into the woods, and live on pinecones and berries. But you can take some vaccine advice along on the way:

1. Think carefully before you exchange your numbers for money. Are you quite certain that you want Google, Apple, and the rest of them to know everything about you, your family, and your health?

2. Don't monitor your number fortune from day to day, whether that concerns health, finances, or social media. That not only leads to increased stress but can also make you more self-absorbed and less moral.

3. Find out for yourself which wine you like most; don't ask an app—at least not an app that needs your DNA to decide.

4. If the numbers on social media become more important to you than the content, remove the apps.

5. If you're over the age of 20, don't post imitations of Charli D'Amelio's TikTok videos. That just looks silly. And you're not going to get rich that way.

As if numbers' creating a new sort of capitalism that influences us both as individuals and as a society weren't daunting enough, we also need to ask ourselves if the numbers influence how we interpret and own the truth too. It's time to dig a little more into that and return to the questions we brought up earlier about how numbers can influence our trust and our empathy.

NUMBERS AND THE TRUTH

Sweden has the highest incidence of rape in the world. Or at least that's what could be read on news screens at the international airport in Istanbul one Friday morning in August 2016. Before the day was over, it had made headlines all over the world, from Sweden to Australia, and had been picked up by both the BBC and Reuters. There was speculation that the news was strategically placed at the airport to deliberately spread the message around the world; the timing was suspiciously close to the critical public statement that the Swedish minister of foreign affairs had made five days earlier about Turkey's having legislated that sex with minors should not automatically be classified as rape. And arguably the news would never have had such an international impact if it had not also contained numbers.

The numbers, from Sweden's own rape statistics, were compared alongside other countries' numbers, which sure enough were lower. That this could be because Sweden has stricter legislation and because rapes there more often lead to both reporting and conviction, as Swedish experts argued, did not get the same traction. There were, of course, no numbers on what percentage of *all* rapes were actually reported.

The numbers from the rape statistics, on the other hand, took hold. The following year they gave rise to news headlines around the world again, which asked, "Is Sweden the rape capital of the world?" (the headline writers didn't seem to be aware that Sweden is a country and not a city, perhaps because there are no numbers on that...). This time it was a British member of the European Parliament who deployed the numbers in a debate on admitting asylum seekers, by pointing at how the number of reported rapes in Sweden had increased in recent years in keeping with the country having taken in more refugees. In this period, Sweden had broadened the definition for rape—but the British politician didn't take that into account either. There were, of course, no figures on how many rapes, according to the new definition, had taken place before it was adopted, so the world could only know for sure that Sweden had the highest (and for several years a skyrocketing) number of *reported* rapes.

The first time I was really struck by the fact that numbers direct our attention was when I wrote my book *Monster* and tried to understand why the United States has the most serial killers in the world. Why not, for example, China or India, each of which has a much larger population? And why did Russia, which is also a very large country, have almost no serial killers at all?

There were several conceivable explanations, such as, for example, that the United States had more TV violence, but when I tried to compare between the countries, it wasn't possible. I found no numbers on serial killers in the other countries. It was only possible to find numbers for the United States, and they were of course highest in the world. When I tried to see if there was a connection between an increase in TV violence and a greater number of serial killers, I couldn't go back farther than the 1970s. Before that, you see, there were no serial killers in the United States because they hadn't come up with the definition yet.

Micael

THE ONLY TRUTH WE NEED?

Numbers are hard to argue against, even when they aren't the whole truth, because they are the part of the truth we trust. People can think differently about what "most" or "a lot" signifies, and we don't need to agree, but numbers are the same for everyone. Even if they aren't the whole truth but just part of it, they become the only truth.

Maybe those rape headlines ring a bell; perhaps you even remember some of the numbers. But do you remember where the numbers came from? Surely the source doesn't matter that much anyway, because the numbers are the same for everyone, regardless of where they come from, aren't they?

What would you say if we told you that the numbers in reality come from a fictitious research institute? You might think we were messing with you, and you'd be right (sorry, we couldn't help it): the numbers come from the Swedish Crime Prevention Board, which is extremely real and reliable. But you weren't sure for a moment, were you?

Studies show that when people read news articles *without* numbers, they judge the credibility of the statements based on the source. However, when the articles contain numbers, the source plays almost no role at all. We see what other people say, believe, and think as only their version of truth, but we see numbers as unquestionable—as the only truth we need.

A tangible and slightly disturbing example of this is a study in which researchers had people read two news articles about the victims in an earthquake disaster in Indonesia, one with statistics and one without. The researchers measured the people's eye movements and determined that those who read the version with statistics looked less at the pictures of the disaster and the victims (and therefore also responded with lower amounts to the question of how much they were willing to donate to help people who were affected).

The risk is that numbers make us think with less depth. That would explain the result from a brain scan study where the research subjects listened to news in versions with and without numbers. It turned out that the prefrontal cortex was activated less in research subjects who heard the news with numbers included. The prefrontal cortex is that part of the brain that controls empathy and our capacity to shift perspectives and change viewpoints. The researchers went so far in their conclusions as to write that the numbers deactivated brain activity in the trial subjects.

Something similar seems to happen with those who write the news. An American content analysis of over 100,000 news articles and posts in social media showed that the larger the numbers journalists reported on, the fewer and weaker the emotional expressions they used. The numbers seemed to have a numbing effect. The bigger the numbers, the less need for a personal perspective of one's own.

In the number epidemic we find ourselves in, this can have major consequences. Especially as mass media researchers have found that news with numbers is given more space and that journalists prefer to report on news that involves numbers almost regardless of what those numbers are. The researchers call this the number paradox: journalists feel less need to verify the truth content in numbers because they assume numbers are always verifiable and therefore can be checked by anyone. The paradoxical conclusion they come to, therefore, is that the numbers are true.

But as we're sure you can guess, the numbers aren't always true.

FAKE NUMBERS, REAL NEWS

It's possible to make up numbers—for example, that this book has sold five million copies worldwide (it hasn't—yet—but it was both easy and a little encouraging for us to make that up). Or that the monster car Hummer only costs $1.95 per mile in fuel compared with a Prius, which costs $3.25 per mile, as news headlines maintained in 2007 (despite this only being true if you divided the cost of driving the Hummer by 35 years of driving, against 12 years for the Prius, as the PR agency behind the numbers did). Or the numbers and statistics the Pentagon fed to reporters during the Vietnam War about body counts and numbers of weapons captured, to get favorable news headlines and rally popular support for the war.

As a neighboring country of Russia, we Norwegians stayed firmly glued to our TVs and smartphones when Russia invaded Ukraine on February 24, 2022. Soon, news stories about the "Ghost of Kyiv" emerged both in social media and on reputable news channels, and I found them deeply fascinating. The Ghost of Kyiv was the nickname of a Ukrainian

MiG-29 pilot who allegedly won no less than 6 air fights in the skies over Kyiv during the first 30 hours of the invasion. The Ghost shot down two Su-35s, two Su-25s, an Su-27, and an enemy MiG 29. Later, on February 27, the Security Service of Ukraine reported in a Facebook post that the Ghost of Kyiv had shot down 10 aircrafts. In the weeks to come, various news outlets reported that as much as 40 planes had been shot down by this mythical air warrior.

These exact numbers, backed by computer-generated footage of the Ghost of Kyiv winning an air fight, made the story appear precise, credible, and true. The numbers and the video spread like wildfire on social media and were also shared by the official Twitter account of the Armed Forces of Ukraine. Later, Ukraine's Air Force Command admitted that the Ghost of Kyiv was a superhero legend, and it turned out that the viral video was made by a YouTuber and adopted from the 2013 video game *Digital Combat Simulator.*

Helge

It used to be called propaganda; today it's called "fake news." And just as numbers were a classic propaganda trick (google "propaganda" and numbers show up high on the list of tips for how to recognize propaganda or—yes, unfortunately—create your own), they are equally effective in fake news—in two ways.

For one thing, we don't even need to believe that the

numbers are true to be influenced by them anyway. Take shootings in Malmö, Sweden, for example, a city with just over 300,000 inhabitants. Which number do you think seems most reasonable:

600 people shot to death per year?

10 people shot to death per year?

If we maintain that the right number is 600 people per year, you're probably going to think that number sounds too high; almost two people aren't shot to death every day there. But if, after having given you these alternatives, we ask you to guess what the actual number is, your guess will quite certainly be higher than if instead we'd asked if you thought that the right number was 0 or 10. Then you probably would have guessed something in that range, not 55 or 78.

The numbers you encounter create a frame of reference, regardless of whether they are true or not. We know that because we've tested it.

When the fatal shootings in Malmö were being reported on most heavily (which, as it happens, was right before the Swedish parliamentary and municipal elections), we divided just over 1,000 randomly chosen Swedes into two groups and let one group react to our statement that 600 people were shot to death, while the other group got to react to the statement that 10 were. The group that saw the higher number thought (thank goodness) that it was too high and must be false. But when they themselves guessed the actual average number, their estimate was almost twice as high as that of the group that had seen the lower number (which was perceived as more credible). The group that had seen the higher number also thought there was greater insecurity and uncertainty in the city. Even though they didn't believe the number, it influenced their view of the truth anyway.

Psychologists call this "anchoring": when forming our own understanding about something, we need a guide to proceed from, to anchor our understanding. And because numbers quickly make their way to the neurons in our brains, we don't have time to defend ourselves before they take root there and influence our assessments, even when we know that they're wrong.

For example, if you're asked whether you think that the probability that a nuclear war will break out in the next 10 years is greater or less than 90 percent, you will probably answer less. And if you're asked whether the probability is greater or less than 1 percent, I suspect you'll answer that it is greater. But when you're then asked to come up with the probability yourself, you will be influenced by the question you had before—you'll guess higher if you got the first question (with 90 percent planted in your brain) than if you got the second. Researchers who conducted that exact experiment got exactly those results—even when they did the experiment again and asked people to think about how nuclear wars arise, and even when they told people that the numbers 90 and 1 had been plucked out of thin air. Again and again those who had seen the higher number guessed that the probability was around 25 percent (!), while those who had seen the lower number put it considerably lower, at around 10 percent.

It gets even wackier when we believe that we've freed ourselves from the incorrect number and so feel more secure about our own estimated number. When students at the school of economics in São Paulo had to estimate the value of large companies on the stock exchange, their guesses were unsurprisingly influenced by whether they first had to decide whether the value was higher or lower than a certain number (far too high or far too low). But they also proved to be considerably more certain of their own (almost equally incorrect) numbers than the students who did not get to see any other numbers before their guesses (which were actually much closer to the truth); they were even prepared to bet money on it.

So the numbers fool us twice: they influence us whether we believe them or not, and when we don't believe them, we become even more certain of our (influenced) view of the truth.

Is there anyone besides me who in the 1990s didn't dare eat anything that contained the sweetener NutraSweet because researchers said it could cause a brain tumor? The story of how hazardous NutraSweet was alleged to be is a fascinating tale of how even completely true numbers can easily lead us astray if they are linked together in new ways. The researchers saw an alarming increase in brain tumors three to four years after NutraSweet was introduced on the market in the early 1980s. They published a study on this in the *Journal of Neuropathology and Experimental Neurology* that got a lot of attention. Even if all the data on which the study was based was correct, the conclusion they drew was completely erroneous. As Charles Seife so brilliantly explains in his book *Proofiness*, a number of other things also increased dramatically during that time period in the 1980s: Sony Walkmans, Tom Cruise posters, shoulder pads, *Donkey Kong* games, and government spending. In fact, there was a stronger link between the sale of Nutra-Sweet and government spending than between NutraSweet and the number of brain tumors. Do you follow? This classic trap—the belief that a correlation between true numbers is the same as cause and effect—has given rise to an incredible number of erroneous articles in the press, false truths, and conspiracy theories.

Helge

As if this weren't enough, the numbers don't even need to have anything to do with the matter. We are number animals who instinctively react to the numbers that are available, regardless of what they are. In some classic experiments, students at the Cornell and Harvard business schools got to predict how many points the fictitious basketball player Stan

Fischer (with jersey number 54 or 94) would make in the next NBA game or how much money they would spend at a dinner at a new, made-up restaurant in town (called Studio 17 or Studio 97). The Stan Fischer with the higher jersey number was expected to make considerably more points in the game, and the restaurant with the higher number in its name increased the students' calculated dinner check.

Here is where it gets really frightening. What if the numbers that pop up all the time in our lives get anchored among the neurons and influence our understanding and decisions about other things that are happening at the same time?

What if the high number on your step counter gets you to take more money out of the ATM? What if the high number of likes on your latest picture on Instagram gets you to make a higher offer than you otherwise would have thought was reasonable on eBay or Redfin?

We got curious. We asked approximately 1,500 individuals to write down how many steps they had taken during the day (the majority have a health app on their phone that automatically counts the steps; those who didn't had to guess to the best of their ability). Then we asked them to put a number on how much they would be prepared to pay for a one-bedroom condominium in their city. And guess what? The higher the step number reported, the higher the price they were prepared to pay for the condo. Now perhaps you're thinking this is because people who live in bigger cities walk more (because of the greater distances) and housing in bigger cities costs more, but we controlled for that. Higher step numbers yielded willingness to pay higher prices regardless of city. It's also possible that people who took more steps felt capable and "rewarded" themselves by being prepared to pay more for the condo. But the effect was the same when we asked participants to *guess* the average price of a one-bedroom apartment in their city.

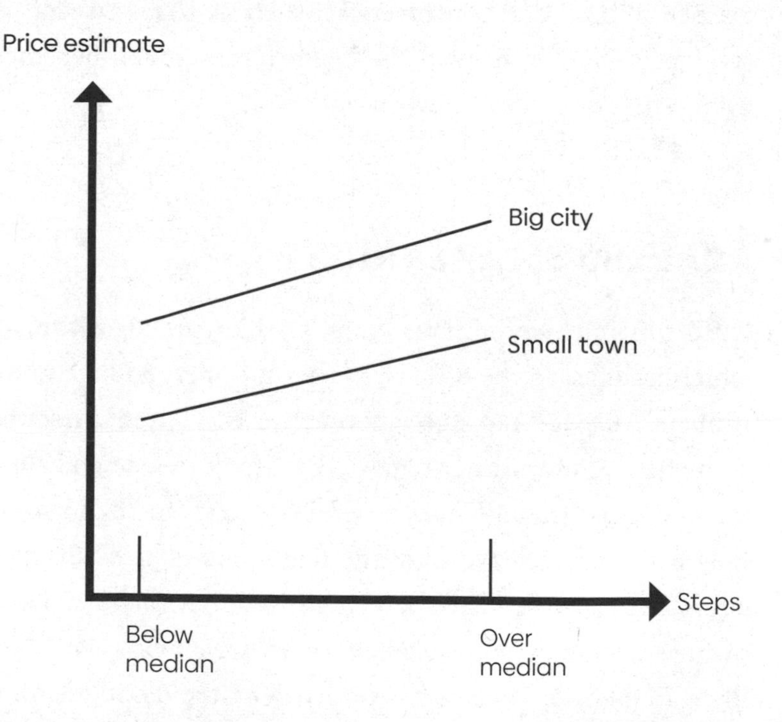

What if algorithms could sense the size of the numbers you are surrounded by at any given moment and adapt to them in fake news or in advertising (now we're getting a little Orwellian)?

In a way this already happens. Social media algorithms react to numbers in the form of views, comments, and shares and give more space to the posts that get a lot of those goods. And we have already determined that news with numbers creates more clicks, generating a sort of double number effect: the numbers in the posts generate more clicks, which in turn get the algorithms to spread the posts even more. If the numbers are sensational and controversial—such as that "Sweden is the rape capital of the world"—the push presumably becomes even larger, a real fake-news payday.

REAL LIKES, FAKE TRUST

Unfortunately, we humans function like the algorithms—which leads us to the other way that numbers in fake news influence us. It's not enough that we can't free ourselves from the numbers in the news; we also can't defend ourselves against the numbers around the news in the form of how many people have seen and liked a news piece. Studies show that people perceive news items that have attracted lots of likes online as more credible than news items with few likes. In addition, people have a harder time distinguishing which news items are true or fake when they have many likes—it's as if the high numbers get in the way of critical thinking. When the true and false news items have few likes, people don't have the same problem distinguishing them.

More ridiculous is how commonly people like and comment on posts without actually clicking on or reading them, meaning the likes we are influenced by don't even need to be genuine. In fact, the number doesn't even need to apply to likes in order for us to be influenced by it; views are enough.

We did an experiment in which people got to see a positive or negative post about a fictitious person, with a number that said the post had been viewed either 20 times or 2,000 times. Those who got to see the positive post perceived the fictitious person more positively when they saw that it had been viewed 2,000 times versus a mere 20 times. In the same way, those who saw the negative post thought more negatively of the person when they saw that it had been viewed 100 times more. Despite that, those who saw the high number were just as certain as those who had seen the low number that they had *not* been influenced by how many other people had seen the post (on the contrary they were more inclined to believe that *others* had been influenced by the number).

For the past few years, it's been possible to see how many times an academic article has been cited by other researchers. The intention is good, to allow researchers to find out which articles have "informed" (as it is so nicely called) and made an important contribution to continued research. That number reinforces itself—the higher it is, the more researchers choose to read (and then themselves cite) the article, and so it rises further. The number of citations is even calculated when researchers apply for positions and are promoted, as a measure of how important and strong their research is. This makes me ambivalent, to say the least, about the fact that one of my most cited articles is one I was invited to write about how advertisements ought to be redefined, which many other researchers thought was much too radical and so reacted to in their own articles. One reason that this article has such a high number is because other researchers *don't* agree!

Micael

We can probably thank the intraparietal sulcus (IPS)—that part of the brain housing the number neurons that connect quantities with our primitive survival instincts, drawing us to friendly people and warning us off the unfriendly—for our inclination to be influenced by how many other people have seen a video we're viewing. The IPS also controls our interpretations of other people's intentions. The numbers "translate" others' behavior into a form of collective opinion, for or against, that we ought to join up with or watch out for. But others' behavior doesn't need to mean anything at all—in this case, people have seen the post, period. Perhaps they weren't particularly attentive, didn't bother to even read the whole post, or didn't form any opinion whatsoever. They may even have formed the *opposite* opinion!

For us, the numbers—including numbers that are irrelevant—become acute signals. None of us will likely ever end up in a life-or-death situation where we need to be friends with or flee from 2,000 people or even 20. It's probably good enough to keep track of up to 5 people at a time, as the Pirahã and Munduruku people in the Amazon manage to do without numbers.

And now that we're able to see how "famous" a person is based on how many followers he or she has and how many people look, listen to, and like what he or she does, the risk is that we will think that what the celebrity says is more important and truer in line with the rising numbers. "Many followers equals many truth points." That thought becomes even more distasteful in light of what we noted before, that it's possible to buy social media followers.

In the same way, protests can arise from a kind of number psychosis, where we are dragged along in taking a stand more because so many others seem to think an issue is important than because of what we think ourselves.

> I remember when the "trillingnöt" praline was removed from the Aladdin chocolate sampler. The newspapers reported on the outcry from thousands of people who reacted online, and the manufacturer, who had chosen to remove it from the box because it was much more expensive to manufacture than other pralines, decided to sell it as a completely separate product. But sadly that had to be quickly abandoned when demand turned out to be extremely limited. So much for that outcry.
>
> Micael

We conclude with some number vaccines against truth distortions:

1. Keep in mind the number paradox. Simply because numbers are *verifiable* doesn't mean that they are *verified.*

2. Even if the numbers are true, they are never the *whole* truth.

3. Be careful with numbers; they can reduce people's empathy and in the worst case undermine important messages.

4. Be aware that numbers can get anchored in your head and influence you even when you know that they are unrelated or actually incorrect.

5. Remember that the numbers that surround a message don't say anything about its truth content. That many people have seen it or that the messenger has more followers doesn't make it more important or accurate.

NUMBERS AND SOCIETY

Hang on to the thought of numbers as truth a little longer. Of how numbers "stick" inside us, mislead us, and sometimes are simply wrong. And of how we let ourselves be influenced by numbers, even when we *know* that they're wrong.

We're not done yet.

So far in this book we've talked a lot about how numbers can influence you personally—your self-image, your opinions, your performance and relationships, your motivation and happiness. But what influences us as individuals also, of course, influences us as a group. A book about how we let numbers guide our lives is at the same time a book about how numbers influence society in general. The societal perspective is there all the time if we just look around a little.

Society is governed by numbers. And what almost *always* gets the last word when corporate executives, judges, politicians, and bureaucrats make decisions? Yes, numbers. Numbers that are often wrong or misinterpreted. Numbers that are irrelevant or random. Or that tell the story that you *want* them to tell.

Let's take a well-known example from the British 2015 general election campaign. Prime Minister David Cameron asserted that 94 percent of households were better off thanks to the most recent tax changes, Ed Balls from the Labour Party maintained that families with children had paid £1,800 more in higher value-added tax, while Deputy Prime Minister Nick Clegg proudly proclaimed that 27 million people paid £825 less in income tax. Who was wrong? *None of them.* The fact is that all three were right, based on their own, very selective use of numbers and statistics.

Or take the debate on crime statistics in the United States in 2016. Donald Trump re-tweeted a graphic falsely claiming that 81 percent of all white murder victims had been killed by "Blacks." (The graphic cited something called the "Crime Statistics Bureau—San Francisco"). That number was naturally dramatic and sensational, especially considering the FBI's own crime statistics, which showed the exact opposite—namely, that 80 percent of all white murder victims were killed by other white people. That didn't stop the numbers from spreading like wildfire. Trump himself said to Bill O'Reilly on Fox News when confronted with the fact that the number seemed to be 100 percent nonsense, "Hey Bill, Bill, am I gonna check every statistic? I get millions and millions of people, @RealDonaldTrump, by the way."

NUMBERS STICK FIRMLY

One of the innate characteristics of numbers is that they stick firmly in the brain like they've been superglued. And certain numbers lodge in your memory and remain there forever. If you're over 30, there's a good chance that you still remember the telephone number of your childhood home. Or the license plate number on your first car. Certain numbers sail into the brain and anchor; others glide past unnoticed. Numbers also sneak unnoticed into the judgments you make every day. They become a yardstick, whether you want them to or not.

We just wrote about how the numbers we see or hear, like the number of shooting deaths per year, influence our judgments. And there are many similar examples. How much does an adult giraffe weigh, do you think? Assuming you're not a giraffe expert, which very few people are, there's a high probability that you'll need to take a wild guess. And if we give you a clue or reference point, it's fairly certain that you'll adapt to that. If we ask whether you think that a giraffe weighs more or less than 2,000 pounds and then ask you to guess the weight exactly, you will presumably respond with a rather high number. But if we ask whether you think that a giraffe weighs more or less than 600 pounds and then ask you to guess the weight exactly, you will probably answer with a much lower number.

Regardless of whether this concerns the number of gunshots in Malmö, the risk of nuclear war, the value of companies traded on the stock exchange, or the weight of a giraffe, we thus lean constantly toward the "anchor point." And regardless of whether such anchors are false or true, conscious or unconscious, they influence the decisions we make every day.

Amos Tversky and Daniel Kahneman (who later won the Nobel Memorial Prize in Economic Sciences) were among the first to study this phenomenon. In one of their studies, the participants first saw a roulette wheel that stopped at either 10 or 65. After that, the participants were asked to guess what percentage of the countries in the United Nations were African nations. The participants who had seen the roulette wheel stop at 10 answered with low numbers (25 percent on average), while the participants who saw the roulette wheel stop at 65 on average guessed that 45 percent of UN nations were African. The participants were therefore influenced to an unexpected extent by a completely random number that had nothing whatsoever to do with the relevant assessment. The numbers sneak into our brains anyway.

You may be wondering what possible significance this has. The fact that numbers influence our estimate of the weight of a giraffe (around 1,500 to 2,500 pounds) or the percentage of African members of the United Nations (28 percent) is hardly a major social problem, is it?

No, maybe not.

But consider whether the number that has lodged firmly in your brain is the maximum number of immigrants that your country can handle? Or the expected mortgage interest rate in the next 10 years? Or the number of years a criminal should be in prison? Then it suddenly gets a little messier, doesn't it?

And we do know a fair amount about how anchoring functions here too. A series of studies show how a number that is presented early on in a trial, whether it concerns recommended punishment or proposed compensation, systematically influences both jury and judge. If the number is low, the number of years in prison is often lower. If the number is high, the defendant risks a longer prison sentence. In the absence of other information, we humans use numbers as anchors and reference points. And it's proven that we have great difficulty navigating away from that anchor once it's gotten a foothold in our memory.

The same applies to the numbers politicians are exposed to and, not least, the numbers they serve up to voters. They stick.

Research has documented, for example, that anchoring influences how experts arrive at economic estimates, key ratios, and future prospects. Professional estimates of macroeconomic units—interest rates, currency exchange rates, expected economic growth—are extremely important for both politicians and private decision makers. And if these estimates are influenced by relevant and irrelevant numbers, we risk having politicians make poor decisions.

Also, it's a little troubling that the numbers politicians in turn serve us, whether true or untrue, both consciously and unconsciously change our assessments. Just think about all of Trump's fantasy numbers and how they influenced American voters. A Swiss study from 2019 showed how people's willingness to accept a high or low number of immigrants systematically varied according to the different anchor numbers they were served. The anchoring effect was so strong that it didn't matter which political party made use of the number. It influenced individuals anyway.

Several studies have shown that number anchoring is a rather robust phenomenon that influences the results concerning everything from economic and political decisions to people's estimates of Mahatma Gandhi's age, the average duration of intercourse, and the freezing point of vodka.

And regardless of whether the anchor comes from other people ("the broker said that a similar house goes for $500,000") or yourself ("that house is worth at least

$600,000"), it influences your assessments. If you want to sell your TV for $1,000, advertise that it cost you $2,900, and it will seem like a good deal. If you need to borrow $100 from someone, ask for $500—then $100 won't seem like a lot after you're first given a gentle no.

Did you know, by the way, that your personality also determines to what degree anchored numbers upset your assessments? If you're generally a compliant person, you are influenced more by reference numbers. If you're more inclined to question authority, the reference numbers guide your decisions to a lesser extent. But regardless of how we're constituted, the numbers stick firmly in our brains anyway and influence our decisions more than we want to believe.

COMPLETELY TAKEN IN BY THE NUMBERS

Numbers are concrete, exact, and clear, right? We've heard that. Or thought so ourselves.

> Numbers don't lie.
> They are honest, controllable, and neutral.
> A rational and enlightened society is based on
> numbers, not on emotions or opinions.
> We should base our decisions on numbers and facts.
> We are living in an enlightened democracy after all.

The thing is, though, that numbers often fool us. And they lead us to fool one another. In political debates, the one who brings in a number often wins the discussion—has the last word. You don't mess around with a number—in any event, not if the number comes from a national statistical organization, a research report, or the public. Then it has to be true.

Is that correct? Decision makers and politicians (and you yourself) can be fooled by numbers, or fool one another with them, in several ways. Let's look a bit more closely at the two most important.

THE NUMBERS ARE SIMPLY WRONG

The first and most obvious number problem in society is naturally if the numbers are wrong. And there are several reasons (some funnier than others) why numbers can be incorrect or misleading.

PEOPLE LIE

Not always, and not always consciously, but people do tell white lies and censor and embroider the truth a little. For example, in opinion polls—perhaps especially in opinion surveys on sensitive subjects like politics or sex. In a British study that gathered data from heterosexual people from

2010 to 2012, men on average reported that they'd had sex with 7 women, while the women reported approximately half that number. That seems impossible; those extra women must come from somewhere.

Here you may suspect that the men and/or women have embellished the truth a little. A study conducted seven years earlier, in 2003, illustrated this in an elegant way. In that case, people were asked about their sexual habits, and half of participants were connected to a (fake) "lie detector." The result? The number of sexual partners for women increased from 2.6 on average to 4.4, an increase of 70 percent. People often bend the truth a little when answering surveys due to what researchers call "social desirability bias"—that is, we tend to answer questions in a manner that we believe will be viewed favorably by others. We even do this when answering anonymous questionnaires. On topics ranging from political orientation, religion, and immigration to questions about income, grades, health, substance abuse, and the use of contraceptives, we tend to underreport "undesirable" behavior and overreport our good behavior and attitudes. No wonder pollsters often get it wrong.

THERE ARE SYSTEMATIC ERRORS IN THE NUMBERS

Most people remember how the media and the opinion pollsters almost unanimously predicted, the day before the presidential election in 2016, that Hillary Clinton would defeat Donald Trump. A poor professor at Princeton, Sam Wang, was so convinced ("99 percent") that he claimed he would eat an insect if Trump won—which led him to chew a cricket live on CNN a few days later. The cricket tasted "mostly honey-ish, a little nutty," he reported.

Errors arise in opinion surveys for many different reasons: mistakes in the sampling, too small a sample, large margins of error, or simply the wrong question asked. The same question posed in *almost* identical ways can produce dramatically different results. In the early 1990s, for example, CNN (in collaboration with the Gallup Institute) reported that 55 percent of Americans were *against* the bombing of Serbian forces in Bosnia. The same day, ABC News reported that 65 percent supported the bombing. The only difference in the formulation of the question in the ABC News survey was that people were asked if the United States, "along with its allies in Europe," ought to bomb. CNN only mentioned the United States in the question. Likewise the use of value-laden words, such as "pro-life" instead of "antiabortion," naturally produces dramatically different numbers for the same phenomenon. Even a simple yes/no choice can be framed in two fundamentally different ways, producing very different numbers. For instance, in organ-donation decisions, one can either give people an opt-in option ("Check here if you consent to donating your organs") or an opt-out option ("Check here if you do *not* consent to donating your organs"). Same choice, framed in two different ways: you are still free to choose whatever you like. The opt-out option, however, often leads to consent-rates *twice as high* as the opt-in option.

THE NUMBERS IN OUR DATABASES ARE CODED WRONG

Numbers can also make computers go nuts. Remember Y2K, aka the Millennium bug? As the new millennium approached, programmers realized that computers might interpret "00" not as 2000 but rather as 1900, which would be slightly unfortunate, of course. Not only for banks (which could end up calculating interest for minus 100 years), but for every system that depends on correct dates, including airlines, the military, and power plants. Depending on whom you ask, fixing the Y2K bug cost between $100 billion and $600 billion worldwide, so 2000 is probably the most expensive number in history. Luckily, it all went pretty smoothly, except for a small radiation equipment failure at a nuclear energy plant in Ishikawa, Japan.

People make mistakes too. Clumsy fingers and programming errors can produce both small and large effects, random as well as systematic. Individuals can be registered with the wrong income, mailing address, or credit rating, which in some cases can be really quite unfortunate for the person involved. In other cases, wrong coding can be much more dramatic, such as when the health register in Great Britain showed 17,000 pregnant *men* during the 2009–2010 period. At last, fortunately, a little bell rang for someone, and the erroneous coding was corrected.

THE PRECISION OF NUMBERS IS EXAGGERATED

There is considerable uncertainty associated with the great majority of numbers we use every day and with the numbers used by politicians and decision makers, not to mention economists and financial analysts. This uncertainty results from measurement mistakes, statistical margins of error, and, well, the fact that these numbers are often just *estimates* based on uncertain data. We don't *know* what future interest rates, housing prices, or electricity costs will be. Still, these numbers exist because we have prices, estimates, analysts, artificial intelligence, markets, and even futures markets. And the more digits and decimals we add to an uncertain number, the more certain and exact it appears. That the average interest rate for a house mortgage in the United Kingdom will be 3.15 percent in 2027 sounds pretty precise, doesn't it? But consider for a moment the absurdity of including two decimals for this very uncertain prediction. Not all people catch this. When confronted with such overly exact predictions, we risk making decisions too confidently. In the midst of the pandemic, *OECD Employment Outlook 2020: Worker Security and the COVID-19 Crisis* predicted an unemployment rate "at or above the peak level observed during the global financial crisis, reaching 7.7 percent by the end of 2021 without a second wave (and 8.9 percent in case of a second wave)." The answer, just 18 months later, turned out to be closer to half that number, even with several new waves of the pandemic.

The future is very uncertain and often moves in directions we never imagined. Of 11 predictions made by the French commercial artist Jean-Marc Côtè at the 1900 Paris world's fair, the "Exposition Universelle," only three proved on the money. The predictions that turned out to be quite wrong include that we would "domesticate whales and use them for transportation" and that firefighters would "fly around with batwings on." Many years later, in 1964, the RAND Corporation declared that it expected humans to have animal employees by the year 2020. The RAND guys are no idiots—they contributed to the space program and, later, to the development of the internet—but even very clever people often make predictions that are exact but wrong. Making precise predictions about moving targets, such as technology, is particularly risky. This quote from the March 1949 issue of *Popular Mechanics* may serve as a prime example: "Where a calculator like ENIAC [the world's first digital computer] today is equipped with 18,000 vacuum tubes and weighs 30 tons, computers in the future may have only 1000 vacuum tubes and perhaps weigh only 1.5 tons."

That's just it. When something is extremely uncertain,

perhaps we shouldn't attach a very precise prediction or number to it. And that applies to all numbers and quantities, not just estimates about the future. You see, 66.67 percent of all Scandinavians do *not* eat whale just because 2 out of 3 of your Norwegian friends do. The thinner, poorer, and more skewed the data, the less exact should your estimate be. For instance, 80 percent of dentists do *not* recommend Colgate, even if the famous commercial claimed as much. It was later found that the dentists in the study were allowed to name more than one brand, so the majority of dentists named several. And the normal body temperature is *not* exactly 98.6 degrees Fahrenheit. It varies among individuals, throughout the day, depending on the weather and during the menstrual cycle, by as much as 1 to 2 degrees—not to mention depending on how you measure it. You might in fact be perfectly healthy and still fit for a day at the office even if your temperature reads 100.2 degrees Fahrenheit.

NUMBERS MAY BE BASED ON STUDIES THAT HAVE METHODOLOGICAL ERRORS

There are good studies and poor studies. The latter are called "bad science." Weak studies can sneak into even reputable scientific journals, either because of carelessness and poor methodology or because of cheating. Fortunately, they are usually exposed and withdrawn. One such study in the esteemed medical journal *The Lancet*, conducted by physician

Andrew Wakefield and his colleagues, connected autism to the measles-mumps-rubella (MMR) vaccine. This turned out not to be true, and the article was withdrawn. Wakefield even lost the right to practice medicine in Great Britain.

Yet the numbers and the "link" between autism and MMR have stuck in the heads of certain people (especially vaccine opponents), and the results were enveloped in a kind of aura of truth. No smoke without fire, you know. And suddenly, to some it doesn't matter at all that a major study in 2019 including 650,000 children clearly and unambiguously rejected all connections between the MMR vaccine and autism.

About 20 years ago I was at a pleasant garden party far out on the prairie in Illinois in honor of a faculty member, who had just been promoted to professor at the university there. I was a rather young, inexperienced graduate student at the time and remember this professor as a very clever, ambitious, and self-confident man. He was extremely interested in research on food and nutrition and became widely known in part for studies on the size of portions and plates. He later got important public assignments in the United States linked to nutrition and was regularly quoted in the *New York Times* and elsewhere.

The problem was that something was not quite right with the numbers on which several of his studies were based. To be more exact, he forced his data to cooperate—something we call p-hacking. In an email that was later leaked to the press, he wrote this in response to a research assistant who hadn't found any interesting findings in a study that they had done together. "I don't think I've ever done a study where data 'came out' the first time I looked at it.... Think of all the different ways you can cut the data and analyze subsets of it to see when this relationship holds." The professor was here encouraging good old research fudging—that is, searching high and low for chance connections that can later be presented as something new and interesting. If you search long enough for numbers that support your theory, you'll find them at last. It is a bit unfortunate, however, when private and public institutions base their policies and decisions on such p-hacked studies.

Helge

While the professor has always maintained there wasn't any fraud, this little anecdote transports us right into the second main reason that numbers fool us: misinterpretation.

THE NUMBERS ARE MISINTERPRETED

Sometimes the numbers are correct after all, but they are misinterpreted so that the conclusions and decisions based on them are completely crazy. This could be because you see the patterns and connections you *want* to see or because you genuinely misinterpret the numbers and the reasons for them. The combination of the two traps can be powerful.

Let's take the latter first: that you misinterpret a connection between two numbers or two units as a causal relationship. This business of confusing correlation and causality is the favorite subject of many academics, and hilarious examples are happily brought out in the wee hours of a dinner party. A recent Facebook post in the group Friends of Norwegian Agriculture can serve as illustration. The post reads like this: "We constantly hear how damaging it is to our health to eat as much meat as we do in Norway. This image [attached to the post] shows that life span has increased at the same time as meat consumption has increased."

Chew on that a moment while we consider a few other examples.

In 1999, CNN reported, based on a study in the prestigious journal *Nature*, that children who sleep with the light on are much more likely to develop myopia later in life. The numbers underlying the study were crystal clear: sleeping with the light on causes shortsightedness. However, after a little while, other researchers dug into the matter and found a strong link between parental myopia and the development of child myopia and also noticed that myopic parents more often left the light on in their children's bedrooms at night. You see? Parental myopia causes both child myopia and leaving the light on. As one of the researchers dryly remarked, "We think this may be due to the parent's own poor eyesight." Perhaps genetics is a more important predictor of myopia than a night-light after all.

For some topics on which people have strong convictions about causality, it is easy to find supporting correlations everywhere. The tobacco industry relied on dubious correlational data for decades to deny the relationship between tobacco and cancer. And antivaccine and conspiracy sites tend to find overwhelming evidence for claims such as that vaccines cause miscarriages among women. But they tend to overlook the fact that common things occur together commonly. Given the high number of pregnant women receiving vaccines, and the general high number of spontaneous miscarriages, it is only natural to expect that a large number of women, by chance alone, will miscarry within 24 hours of receiving a vaccine. And in case you're still wondering, robust scientific studies have found it to be perfectly safe to take vaccines during pregnancy.

Have you had a moment to think about the Facebook post about life span and meat consumption? Can you think of other numbers that have risen from 1950 to 2020 besides meat consumption and life span? And is it theoretically possible that meat consumption *could* have negative health consequences, even if average life expectancy has gone up?

The difference between correlation and causality is easy to joke about. For example, the consumption of cheese in the United States is strongly correlated with the number of people who died because they got trapped in their own bedsheets. But is there a causal connection between the two numbers? It hardly seems likely.

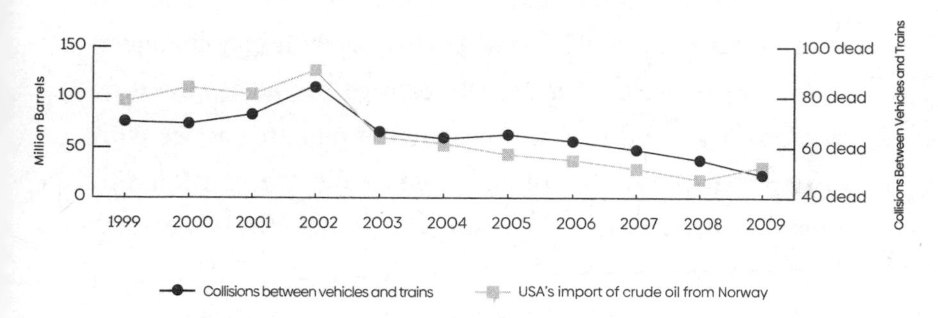

However, in other connections, when two numbers are logically interwoven and covary, it's easy even for veteran researchers to be fooled and assume a causal connection where one doesn't necessarily exist. That happens every day in companies, organizations, political debates, and thousands of homes during dinner-table discussions of various subjects—topics such as abortion, vaccines, economics, dietary supplements, and meat consumption.

We humans also have a tendency to read numbers based on our own values and political opinions. Sometimes we read numbers the way the Devil reads the Bible. And it really isn't so strange that we often *want* numbers to tell us something else. Psychologists often talk about the two related phenomena of "confirmation bias" and "motivated reasoning": we tend to search for and even put more emphasis on numbers and findings that confirm our own viewpoint. If you like wine, you attribute more importance to research that says that wine is beneficial than to research that says it's not. You have no desire to click on articles that say wine causes cancer. If you are skeptical that climate change has been caused by humans, then you read numbers about climate change with different glasses than someone who fully grasps the truth about climate change. If you support Trump, everything that CNN says is fake news.

Anyone can fall into that trap. If you're a Norwegian cattle farmer, it might be that you read the numbers about meat consumption and life expectancy in a different light than a vegetarian. Even "number nerds" fall into the confirmation trap, but perhaps in a different way than you might think. A study from 2017 showed that people with strong mathematical skill more often used it to interpret numbers and problems that *conflicted* with their own worldview. Does this seem counterintuitive? They devoted much less energy to the numbers that confirmed their own viewpoint; those they accepted uncritically. The number nerds thus prioritized dissecting the numbers of "opponents" before critically assessing the numbers that supported their own viewpoint. This too supports the idea that people reason selectively and fall into the confirmation trap.

What about politicians, corporate executives, or your boss? Do you think that they too sometimes fall into the confirmation trap, selectively choosing numbers that support their viewpoint? Or occasionally drawing a completely different conclusion than others, based on the same numbers? Asserting cause and effect where strictly speaking there is only covariation? Or sometimes supporting themselves with numbers that are simply wrong?

Yes. If they are human, they do that. At the same time, more and more decisions in workplaces and organizations are based on numbers and new ways of measuring.

So let's look a little more at how we in society now measure and quantify one another.

NUMBER MARATHON

The year is 1924 and the place is Cicero, Illinois. Employees of the Western Electric Company's Hawthorne factory are on their way to work and to participate in something that will turn into an almost eight-year, much-discussed study of productivity. The purpose of the study is to find out how changes in the physical work environment and surroundings influence workers' productivity. Systematically the researchers start to adjust the workers' surroundings through controlled experiments. First they start with the strength of the lighting in the work space. Some workers have their lighting changed for a period, others don't, and then the researchers measure their productivity. It turns out that productivity increases among those who experience a change in the strength of their surrounding lighting, regardless of how the light is changed. And if that's not enough, productivity also increases in the control group, which has had consistent lighting the whole time! What is astounding and strange about the Hawthorne factory studies is that almost regardless of what variable the researchers change, productivity increases in *both* the experimental group and the control group.

Early textbooks in psychology and organizational behavior call these experiments the "Illumination Studies," and the phenomenon later got the name the "Hawthorne effect." The fact of being observed leads you to change your behavior. And ever since the 1930s, researchers have debated the cause of the results, the methodology used, and whether the Hawthorne effect exists at all. Almost all researchers today agree that being observed or measured can influence everything from people's effort and (short-term) performance to their preferences and priorities.

Since then, measurements and numbers have slipped into every part of our working lives, in private companies, the military, volunteer organizations, and public institutions such as schools, police departments, and healthcare facilities. And with technology always developing, there are more and more numbers, together with even more measuring. We have become so accustomed to it already that we no longer react at all.

It also doesn't help that we are so enthralled by numbers, so delighted by our measuring.

> I remember when I was little, and digital watches became a thing. They showed not only the time but could play different national anthems too—I'm not sure why that was useful, but it felt big. But even bigger were the new timing functions that could count hundredths and lap times. Oh, how we started timing everything! How long were we standing in line in the school cafeteria? How long does it take to eat a meatball (the school had a strict maximum of 10 each)? Or a carrot stick (we ate 10 each to compare, and in contrast to the meatballs, time went up noticeably per lap)? How long is a wink (that required quite a few attempts to get, but I still remember that the average was 19 hundredths of a second)?
>
> Micael

The fact that we can measure most things doesn't mean that we should—or that we always measure the right things. Besides, measuring itself often has consequences. There's a somewhat dystopian but entertaining account in the HBO series *The Wire* of how badly introducing measurements into the public sector can turn out: the police are so obsessed with achieving their quantitative targets that both effectiveness and morale disappear. In real life, teachers who are forced to focus on national tests and measurements find that all other learning is ruined for the students. And politicians set up such unrealistic targets for the police that the targets can only be reached by ignoring real crimes, burying complicated cases, and solving only more trivial ones.

We've already looked at how appropriate and effective measuring is in general. We've seen that measurement can mean that extrinsic motivation suppresses inner motivation and that we risk being turned off something we once really enjoyed. We've seen that a measurement and performance bonus at the workplace can undermine its purpose, and determined that being measured and quantified has unintentional side effects—that we start cheating, become more egoistical, and adapt our behavior to what is actually being measured. This certainly applies to employees and organizations.

AMBULANCES AND PARKING ENFORCEMENT

Employees gladly adapt their effort to the numbers that most easily give rewards, whether this applies to key ratios, response time, customer satisfaction, or low error rates. Companies and organizations do the same. Universities and colleges prioritize the most popular study programs, the scientific journals that give the highest points, and they carefully adapt their strategies according to the measurement parameters in international rankings and accreditations—because their financing from the government is in turn connected to the same measurements. Even hospitals prioritize patients, operations, and interventions based on what yields the best dividend in the point system.

You didn't think that hospitals let themselves be influenced by the tyranny of measurement? Well, in England they even started keeping patients in ambulances to adapt to a new reward system. If patients weren't treated within four hours of arrival, the hospitals were penalized financially. The result? Long queues of ambulances waiting outside until the hospital could be sure to treat patients within the four-hour window. In the United States there were cases where patients were kept alive artificially for 31 days because the hospitals were only paid for patients who survived 30 days after an operation.

The example with the patients makes my own story feel quite trivial. When I was a teenager, I had a summer job at a hamburger restaurant (probably best if I don't reveal which one). Its goal was to become more sustainable and reduce waste, so we would write down every time we threw something away (for example, French fries that had been in the fryer too long or hamburgers that were made incorrectly). This was in the middle of the summer, and they must have had a staffing shortage, because for several weeks I was given responsibility for this policy. I'd barely even started to shave and felt a lot of pressure to get neat, low numbers. My solution? I ate up the old French fries and the mishap burgers, all of it, and discarded nothing. Gorged myself. Fortunately I only needed to be responsible for a couple of weeks.

Micael

You might think that the overall purpose for officials and companies to issue parking tickets is to ensure secure and environmentally sound traffic flow and prevent misunderstandings that have to do with parking. Instead, focus has been directed at the only thing that is easy to measure: number of tickets. The more tickets, the more capable the parking company.

Don't get me started. Many years ago I got a parking ticket for having parked 4.5 meters from an intersection. The ticket was issued right after midnight, and it was the middle of winter. Because it's good therapy to write a complaint, I did so, but of course it wasn't accepted. At last, I made contact with a gentle person at customer service who told me that the minimum was 5 meters but that it was probably hard to see exactly where the curve started because it had been extremely dark outside. The poor parking enforcement officer had to work in the middle of the night. A while into the call she also admitted in confidence that they were a little behind in their budget and therefore had to start a bit of a parking ticket offensive. She advised me to always be extra careful when I parked the car in December.

Helge

MEASURE, COUNT, INTERPRET, IMPROVE

"You can't manage what you can't measure," management guru Peter Drucker famously said.

A challenge with numbers in companies and other organizations, however, is that it's easy to measure and focus on what is simplest to quantify. And much criticism of the wave of new public management, where the public sector is run, quantified, and measured like commercial companies, is simply that a public organization is *not* a commercial company. The public sector is complex, with many considerations and interested parties, and focusing on a number often takes resources and competence from an area that is important for the whole machinery.

Three more or less unspoken assumptions are the basis for a number and measurement culture in an organization. First, it's possible and often necessary to replace experience-based and subjective assessments with standardized numbers and rules. Second, numbers give a predictability and transparency that ensures that the organization is better able to reach its goals. And third, the best way to motivate and control employees is to connect reward and punishment to their performance, in the form of money or reputation.

Considering the challenges that we looked at earlier in the book and that are linked both to people's innate shortcomings and the volatility of numbers, it's not obvious that the three assumptions are always true. The numbers can rather easily mess things up for us; they can also mess things up for organizations and institutions—because of the ways numbers are quantified and measured and how they are used and interpreted to make important decisions.

If you absolutely must measure and count, why not measure something that is fun and motivating? Take a boring number like gross national product (GNP), which most nations use as an indicator of development and progress. What if we instead went with something completely different, something as radical as gross national happiness (GNH). In the young, mountainous nation of Bhutan, GNH has been introduced as the measure of the country's condition instead of GNP. Pretty cool, right?

Although...we have already done that. When we regularly measured people's happiness consistently over a period, we found that each time they got a little *less* happy. Damn it.

Clearly, as a society we have not yet really come to grips with this business of numbers and measurement, so some bits of vaccine advice are absolutely in order in this chapter too:

1. Be critical of numbers. They can be both incorrect and misinterpreted.

2. Be observant of what's called anchoring. The numbers that have stuck in our heads and influence our decisions

can give people longer prison sentences and more expensive houses.

3. Keep in mind this thing about motivated reasoning and confirmation error. Everyone interprets numbers and connections subjectively based on their viewpoints, values, and goals.

4. Numbers lead to comparisons and competition. Think carefully about the areas of life and work in which you want to measure yourself against others. And about who or what you're comparing yourself with.

5. Be careful where you park your car in December.

When we've come this far, the last big question must be this: Do we truly need to measure and quantify everything around us, or is it time to start making the world a bit more mysterious, intangible, and subjective again?

NUMBERS AND YOU

Jesus was not born in the year zero. When he was born, the year was 3761, because time was established according to the Hebrew calendar. His birth year stayed as 3761 for another 500 years, until the monk Dionysius Exiguus decided to count the years starting with the birth of Jesus. But the year of Jesus's birth didn't become zero then either, because zero wasn't invented yet (well, historians think that the first zero was observed in Mesopotamia three years before Jesus was born and then among the Mayans seven years later, but zero didn't come to the Western world until the twelfth century). There is still actually no year zero in our calendar; instead we go directly from –1 (one year BC, "before Christ") to 1 (one year AD, anno Domini, "the year of our Lord"). Jesus was thus born one year after himself!

What does this have to do with me? you may be wondering. The point we want to make is that the numbers we use to calculate time, which are perhaps the most fundamental in our human existence, are invented. What you and Jesus have in common is that someone made up numbers for your existences. He was born first in the year 3761 and then in the year 1, while you were probably born in the mid- to late 1900s or early 2000s. And who knows, in 500 years maybe someone will think of completely new numbers for when you were born.

The same applies to all numbers we use. The numbers that influence our bodies, self-image, performance, relationships, and experiences are *invented*. They are currencies, yardsticks, and truths that we've made up. Some you have perhaps made up yourself, while other people or machines have made up most of the rest. Regardless, they are invented.

Our advice to you is to remind yourself of that sometimes. We've filled this book with both amusing and awful examples of how numbers influence us, most often without our awareness, since we take numbers for granted, without questioning them, thinking that we *are* our numbers. But like all fiction, the numbers don't completely match reality. They have a lot of limitations.

NUMBERS AREN'T ETERNAL

For one thing, *numbers aren't eternal.* They can change at any time. While time has existed for all eternity (or at least for the approximately 14 billion years the universe has existed, as far as we know), the numbers for time have changed a few times. They went from totaling 4,000 years to suddenly being reduced to 500 (approximately) when Dionysius pushed the reset button and invented a new era. And 1,500 years later we fast-forwarded time to billions of years after astronomers came up with ways to date the universe (which has existed a lot longer than the short period since Jesus was born). Soon, perhaps, we'll have thought of new numbers for time again. After all, the 2020 Nobel Prize winner in physics, Roger Penrose, thinks that there must have been another universe *before* our universe, which should probably increase the time factor by another few billion (the difference between the year 3761 and 1 AD suddenly seems trivial). These feel like revolutionary changes of time, to put it mildly, but they're not. All those years have always existed as part of all time. It's just that we've invented new numbers for them.

The year was once only 10 months long (that's why the last month of the year starts with "dec," from *decem*, which means 10 in Latin) until the Romans added another 2 months to bring it in line with the solar year. And in Norway and Sweden in the Middle Ages, every year was one day too long, up until the eighteenth century, when those countries changed from the Julian calendar (itself a reform of an earlier calendar system) with an annual leap day, which always missed the solar year, to the Gregorian calendar, which only has one leap day every four years. To sync up with other parts of the world that had already changed calendars much earlier, we had to shorten the month of February to 17 days for the year we in Scandinavia made the switch.

There are many more examples to show that numbers are not eternal. Look at the sports world. If you were to google a video from the past few years of some gymnast or figure skater who got a 10 in a competition, you might wonder if it's all a visual trick, so unbelievable are their performances. If you then look at some black-and-white film from a 10-point program in an Olympics from the first half of the twentieth century, you could catch yourself thinking, "I could probably do that too with a little coaching." Of course, you almost definitely couldn't, but the point is that gymnastics are a lot more difficult now. Comparing a 10 then with a 10 now is like comparing apples and pears, or perhaps counting sticks and computers, which basically have the same function but differ so much in performance that it's not even relevant to compare them.

Speaking of performance, it was amazing that Sweden took silver in the soccer World Cup in 1958, but being number 2 out of the 16 teams that participated in the championship is still not really the same thing as being number 2 nowadays, when 32 teams participate (or in future championships, when the plan is for 48 teams to take part). And let's not talk about tennis, where the algorithm that ranks the players in the world has changed five times in the past 25 years (considering time periods, which tournaments count and how matches are weighted), so that it's impossible to even compare the players' placement numbers over time.

The numbers you assign yourself aren't eternal either. Take your experiences as examples, you miserable, fussy critics. The 5 you gave as a grade on that movie you saw or the dinner you had a few years ago would probably correspond to a 4 today, or in the worst (grumpiest) case, only a 3. A "5" experience is presumably something quite different for you today.

NUMBERS AREN'T UNIVERSAL

Next you should remind yourself that *numbers aren't universal*. Here we can use time as an example again. The calendar is not in the 2020s for everyone right now. The Hebrew calendar continues to tick along and is already into the 5780s. The Muslim calendar is instead in the 1440s, while the North Korean calendar is far behind. The North Koreans just recently started to count and haven't gotten farther than the 110s (because for them time began when Kim Il Sung was born, something the rest of us missed). And the year when

February was 11 days shorter in Norway and Sweden to catch up with the calendar change didn't even happen in the same year in both countries; it occurred in the year 1700 in Norway and in 1753 in Sweden. Even though we've lived on the same planet in the same universe the whole time, we have completely different numbers for that time.

Another clue that numbers are not universal is the existence of various currencies. Buying the magazine *The Economist* isn't the same wherever you find yourself. You would presumably be prepared to pay more for it in the United States than in Norway or Sweden because the dollar works with lower numbers than kronor. Even though it's exactly the same amount of money that disappears from your account, it hurts less to lose $6 than 50 kronor because we forget that numbers are not universal and instinctively feel that the number 6 is always lesser in value. Researchers call that the "denomination effect," when money (which holds the same value regardless of where you are—you don't get poorer or richer simply because you step over the border into another country) is reduced to numbers in our heads, which we think are universal but are actually different depending on where you are. For that reason, we spend more where the currencies operate with low numbers and less where they're high (even though we have just as much money in both cases).

But you would actually have to pay more for the magazine in the United States than in the Nordic countries because the magazine salespeople there would set the price at $5.99

in an act of so-called psychological price setting (a little like psychological age and its magic boundaries, where we attach greater importance to the first number, remember?). The seller lowers the price by one cent so as to lower the first number from 6 to 5 and get people to experience the magazine as less expensive, while in the Nordic countries they would set the price at 49.50 kronor (which corresponds to five times as big a reduction counted in cents). There are actually studies on *The Economist* (unsurprisingly conducted by economics professors), and for that matter on cornflakes and many other goods, showing that they cost different amounts in different countries even though they ought to be worth exactly the same, precisely because of different magic price points.

A third piece of evidence that numbers are not universal is of course those you assign yourself. It's not enough that you're fussy; you also prefer certain numbers. If you identify as a woman, you're slightly more inclined to grade the hotel with an even number, while, if you identify as a man, you're more inclined to give the exact same hotel that you experience in the exact same way an odd number. Since *everyone* is more inclined to round up even numbers and round down odd numbers, that could explain why women more often assign higher grades.

There are also cultural differences in the numbers we assign. Asians are more inclined than Europeans to choose even numbers and to choose values in the middle instead of closer to the ends of the scale. An "Asian 6" could thus correspond to a "European 7" as a grade and an "Asian 4" could correspond to a "European 3."

NUMBERS AREN'T ALWAYS CORRECT

You would be wise to remind yourself that *numbers aren't always correct*. In any event, not automatically. We've already determined that people tend to *believe* that something becomes true when a number is assigned to it. But even with the best intentions, people—since it's always people who come up with what numbers will be used and how they'll be calculated—can count wrong.

> May I tell one more thing about Jesus? I've tried to restrain my nerdiness so far, but this is of course the last chapter. I dug a little into how that monk, without Google or anything, could be so sure of when Jesus was born, 500 years later. I found no information about that, but on the other hand I discovered that researchers seem to agree that Dionysius must have calculated wrong; they just aren't in agreement about which year Jesus was born instead. Historians think that he must have been born in 4 to 6 BC, because that was when Herod ordered all newborn boys killed (which, according to the Bible, happened in connection with the birth of Jesus, the boy baby they missed in the barn out in the countryside). Astronomers, on the

other hand, think the star of Bethlehem, which shone over that barn, must have actually been the slow-moving comet that passed over Bethlehem in 5 BC or the convergence of Venus and Jupiter that occurred in 2 BC; thus Jesus must have been born in one of those years. So Jesus was literally ahead of his time. The number 1 that we start our calendar with (which really ought to be 0) should actually be –5 or perhaps –2 (or –4 or –6...).

Micael

Most of us are actually fairly bad at counting, especially when it comes to large numbers. As you recall, our brains are constructed to handle sizes and quantities that we encounter in our everyday lives and that are possible to add up in rather small numbers, but because a big number and an extremely big number look almost the same, we mislead ourselves. Or have a hard time grasping them at all. If we say to you that 1,000,000 (one million) seconds corresponds to 13 days, how long would you say that 1,000,000,000 (one billion) seconds is? Presumably you wouldn't think that a billion seconds is 31 full years (which is the correct answer). And we're quite certain that you would think that 1,000,000,000,000 (one trillion) seconds is far less than 31,688 years (which is the correct answer).

The numbers are so large that they become impossible to relate to and differentiate. That is one of the explanations for why people get "speed blind" and take out loans that are too big ($200,000 doesn't feel like that much more than $100,000), gamble away money (losing $100,000 doesn't feel 10 times worse than losing $10,000), or speculate away billions on the stock market (the record—for now—is held by a finance manager at the American bank JP Morgan, who in 2012 did away with $9 billion).

Machines can also mess things up. In 2011 *The Making of a Fly* became the world's most expensive book when the price rose to almost $24 million on Amazon. The book about genetics had, only a day or two earlier, sold for $35 without people exactly chasing their cards down to order it. The explanation turned out to be that two booksellers used the same algorithm to search for book prices from competitors and then offered the same book at 1.3 times the price (buying the book from the competitor if someone ordered it and shipping it to their customer). Every time one of the booksellers' algorithm multiplied the price by 1.3 in its offer, the other bookseller's algorithm reacted by multiplying the price by 1.3 in its offer, back and forth (and it actually didn't need to be multiplied that many times for the price to skyrocket).

Last summer we ran a digital race together, the whole family, where it was okay to run any route we wanted, and then the GPS calculated when we crossed the "finish line" after exactly 10 kilometers. We decided to run together and cross the finish line at the same time—teenage daughter, teenage son, my wife, and I—in a way that you can never do in the crowd in a "normal" race. But when my daughter and I got a message at almost the same time that we only had 200 meters left to the finish line, my son and his mom, who were running right beside us, still had over 600 meters left. It didn't do any wonders for sibling affection that big sister beat little brother by well over a minute and ran a "victory lap around the stadium" while he sprinted his last 400 meters. As to whether this was because we had different phone models, or because my daughter ran "better," as she thought, or something else altogether, we didn't get an answer.

Micael

NUMBERS AREN'T ALWAYS EXACT

It's always wise to keep this in the back of your mind: *numbers aren't always exact*. They feel so precise, with decimals and everything. But even with decimals they are often rounded. Take the constant pi, for example, used to calculate round things like circles and spheres. Everyone knows that pi equals 3.14. But that's not really correct, says someone who knows that pi equals 3.14159265. People who are just a little nerdy stop there, while people in memory competitions continue to rattle off hundreds or even thousands of decimals.

So it's true that pi is 3.14, but not exactly. Which becomes a rather big problem if it concerns setting the course of a rocket ship toward the planet Mars, for instance, and missing by several kilometers or dating the universe and missing by a billion years or so (ask NASA).

In high school I remember that the whole class got a disproportionate scolding for being lax and approximate with rounding numbers in math class. This was the fall of 1991 in Stavanger, Norway, and both the teacher and the whole region were probably a little shaken after various unfortunate calculation and rounding errors. A classmate and his family had actually left the city because of a calculation error. His father had been important in the calculations concerning the concrete foundation for the oil platform Sleipner A. The foundation, which cost 1.8 billion Norwegian kroner, came loose and sank in the fjord with such a powerful crash that it registered as an earthquake at 2.9 on the Richter scale all the way over in Bergen. That sort of thing leaves its mark, especially if you're a somewhat cautious and nervous math teacher.

Things didn't get better when a rounding error killed 28 American soldiers and wounded 100 during the Gulf War that same year. The American Patriot defense system had whittled down the number of decimals to 24 (and then rounded). That may sound like a tiny detail, but our math teacher assured us that it means a heck of a lot if you're trying to localize a launched SCUD missile.

> I learned this in any event: write down enough decimals or write the number as a fraction, otherwise you risk people dying or causing an earthquake.
>
> Helge

In case these rounding examples are a little confusing (we humans are, as discussed, not very good at understanding those kinds of big or extremely small numbers), let's take something you can more easily relate to. Your own numbers. Just because you assign a grade of 4 to a film, it doesn't mean that the film is exactly as good as every other film you have given a 4 to, right? But because the scale only lets you choose 3, 4, and 5, it will be a 4, even though the film perhaps was maybe only a 3.5 (which you round up because even numbers are more beautiful). Two films that you actually feel very differently about, just as different as the difference between a 3 and a 4 or a 4 and a 5, thus get the same grade!

It's just as mad if you take the average of other people's grades. If you're a poor professor, subjected to RateMy Professors.com, and half of all those who assigned grades to you chose the number 3 and the other half chose the number 5, you would receive the average grade 4, which you could read as the majority thinking you're pretty good. However, in reality, half think you're mediocre, and the other half love you (or something like that). Or maybe you're a little "edgy" (a word maybe more applicable to stand-up comics than professors) and get half 1s and half 5s. Anyone who sees the average grade, a 3, will probably presume you are right bang in the middle of all the mediocre stand-up comics or professors and scroll ahead, when actually you must be something quite special to invoke such strong reactions!

It's the same, of course, when you do an average of the grades you yourself assign. The 4 on a restaurant visit is the average of the 5 you give to the food, the 4 you put on the service, and the 3 you put on the cleanliness of the restroom. If you'd left out the restroom or perhaps hadn't visited it the whole evening, the grade would have been closer to a 5 instead. Regardless, that 4 is far from an exact grade of your experience; it conflates all the elements and would be of little use or even misleading for someone who wanted to form an understanding of what they could expect from the restaurant ("everything seems pretty good?"). A gourmet who would love the food might not go there after having seen your average 4 while someone with a sensitive stomach and fear of germs might go there and pass out in the restroom.

NUMBERS AREN'T OBJECTIVE

This brings us to the final point we want to make about these numbers we invent: *numbers aren't objective.*

Okay, 3 pieces of fruit are always 3 pieces of fruit; that number is objective. But if the 3 is instead a grade of how the fruits taste, it immediately becomes subjective, even though the number looks exactly the same. (If we're honest, the 3 that represents the number of fruits could be subjective too, because if one of the pieces of fruit was a tomato, then some would count it as a fruit–it has a peel that protects the seeds–while others, for example those who follow the court ruling in the US Supreme Court from 1893, would count it as a vegetable.)

The number you choose when you grade something is not only influenced by your subjective taste but also by other subjective factors, such as what situation and mood you happen to be in. If you have just found a coin on the ground, you are probably going to grade your own prospects with a higher number (there is actually a study on this, from the days when people used coins, which they might happen to lose on the street). If the sun is shining, you will grade your job with a slightly higher number (yes, there's a study on that too), and if your national team won yesterday's match in the World Cup, you're likely to be more satisfied with your personal finances and the whole country's economy and so rate the government more highly (if you're German that is; most of these studies happen to have been done on Germans in particular). If you're hungry, then you set a higher grade on the food you're eating (not completely unexpected) but a *lower* grade on everything from the film you're watching to the shampoo you use and the shoes you have on your feet (this has been tested in studies, but not, we should maybe point out, at the same time).

NUMBERS ARE (AFTER ALL) AMAZING

So there we have it. Now we've given you some advice to hopefully manage a little better in the number epidemic.

We don't think that the solution is to eradicate numbers, to stop measuring and counting and comparing. We just need to learn to live with these phenomena in a smarter way. Because numbers are completely amazing. We have focused on all the various hazards and risks associated with them, but, as we said in the introduction, we do still love them. In many ways they are the soil for our civilizations. All the great historical civilizations, from the Sumerians to the Romans and the Mayans, developed, and were developed by, their own number systems. And today we are an almost global civilization that is united through a common number system. We may have thousands of different languages around the world, but we all speak the same numbers (we just need to learn how to understand them as well).

Thanks to numbers we can keep track of whether there are more than 5 peanuts in a jar, and we can divide up how many grains we want in however many piles we want. We can conserve, plan, engage in trade, and share anything. Without numbers we wouldn't have the time or the ability to understand the universe (as best we can). Because of numbers humanity will soon be able to investigate new worlds, even before the end of this century, if we believe researchers.

Numbers can help us do basically everything, and that's important to remember. They're there to help. That's why they were invented to start with. Because all those numbers that sneak into your performance, relationships, and experiences, which influence your self-image and even your body, are *invented*. They only exist because someone, at some time, thought they would make life easier for you in all that you do. But the numbers are only helpful so long as you keep in mind that they actually aren't eternal: they change and take on other meanings sometimes. They shouldn't be used to make comparisons over time or to use as a yardstick for the future (when the same numbers might not even be relevant). Keep in mind that they aren't universal and stop comparing everything and yourself with everyone else. Don't blindly believe in them because they are actually not always correct and exact. And never forget that a great many of the numbers in your life are numbers that you yourself have actually invented.

Sometimes you can even forget about using numbers at all. Tell your friends about your magnificent hotel stay instead of grading it. Write a review of the book with words instead of numbers. Enjoy the restaurant without checking whether your Instagram friends think it's good. Look at yourself in the mirror instead of getting caught up in BMI or the numbers on the scale. Have sex without watching the clock.

And remember,

1. Numbers aren't eternal: beware of comparing them over time and be open to the fact that their significance will change.

2. Numbers aren't universal. Even if they look the same, they can mean different things and have different values that differ between countries, cultures, and people.

3. Keep in mind that numbers are not automatically correct. Both people and machines can count incorrectly, consciously or unconsciously.

4. Even if the numbers are correct, that doesn't mean that they're exact. Almost all numbers are rounded in some way. Watch out for letting them incorrectly restrict your thinking.

5. This is perhaps the most important point of all: numbers are almost always subjective in some sense. They (and you!) become what you make of them. Always use numbers with care and always use your own judgment.

SOURCES

FOREWORD

Becker, J. (2018, November 27). Why we buy more than we need. *Forbes*. www.forbes.com/sites/joshuabecker/2018/11/27/why-we-buy-more-than-we-need/?sh=4ad820836417.

Ford, E. S., Cunningham, T. J. & Croft, J. B. (2015). Trends in self-reported sleep duration among US adults from 1985 to 2012. *SLEEP, 38*(5), 829–832.

Larsen, T. & Røyrvik, E. A. (2017). *Trangen til å telle: Objektivering, måling og standardisering som samfunnspraksis*. Oslo: Scandinavian Academic Press.

Mau, S. (2019). *The metric society: On the quantification of the social*. Medford, MA: Polity Press.

Muller, J. Z. (2018). *The tyranny of metrics*. Princeton, NJ: Princeton University Press.

Nurmilaakso, T. (2017). Prisma Studio: Pärjääkö ihminen muutaman tunnin yöunilla? *Yle, TV1*. https://yle.fi/aihe/artikkeli/2017/01/31/prisma-studio-parjaako-ihminen-muutaman-tunnin-younilla.

OECD (2009). *Society at a glance 2009: OECD social indicators*. Paris: OECD Publishing.

Seife, C. (2010). *Proofiness: How you're being fooled by the numbers*. New York: Penguin Books.

SVT (2018, November 12). Stark trend—svenskar byter jobb som aldrig förr. *SVT Nyheter.* www.svt.se/nyheter/lokalt/vasterbotten/vi-byter-jobb-allt-oftare.

SVT (2018, July 3). Ungdomar sover för lite. *SVT Nyheter.* www.svt.se/nyheter/lokalt/vast/somnbrist.

US Bureau of Labor Statistics. (2021, August 31). Number of jobs, labor market experience, and earnings growth: Results from a national longitudinal survey. *BLS.* www.bls.gov/news.release/nlsoy.htm.

1: THE HISTORY OF NUMBERS

Bellos, A. (2014, April 8). "Seven" triumphs in poll to discover world's favorite number. Alex Bellos's Adventures in Numberland. *The Guardian.* www.theguardian.com/science/alexs-adventures-in-numberland/2014/apr/08/seven-worlds-favourite-number-online-survey.

Boissoneault, L. (2017, March 13). How humans invented numbers—and how numbers reshaped our world. *Smithsonian Magazine.* www.smithsonianmag.com/innovation/how-humans-invented-numbersand-how-numbers-reshaped-our-world-180962485.

Dr. Y (2019, May 17). The Lebombo bone: The oldest mathematical artifact in the world. *African Heritage.* https://afrolegends.com/2019/05/17/the-lebombo-bone-the-oldest-mathematical-artifact-in-the-world.

Everett, C. (2019). *Numbers and the making of us: Counting and the course of human cultures*. Cambridge, MA: Harvard University Press.

Facts and Details (2018). Pythagoreans: Their strange beliefs, Pythagoras, music and math. *Facts and Details*. https://factsanddetails.com/world/cat56/sub401/entry-6206.html.

Hopper, V. F. (1969). *Medieval number symbolism: Its sources, meaning, and influence on thought and expression*. New York: Cooper Square Publishers.

Huffman, C. (2019, July 31). Pythagoreanism. *Stanford Encyclopedia of Philosophy.* https://plato.stanford.edu/entries/pythagoreanism.

Knott, R. (n.d.). Fibonacci numbers and nature. *Dr. Knott's Web Pages on Mathematics.* www.maths.surrey.ac.uk/hosted-sites/R.Knott/Fibonacci/fibnat.html.

Larsen, T. & Røyrvik, E. A. (2017). *Trangen til å telle: Objektivering, måling og standardisering som samfunnspraksis*. Oslo: Scandinavian Academic Press.

Livio, M. (2002). *The golden ratio: The story of phi, the world's most astonishing number.* New York: Broadway Books.

McCants, G. (2005). *Glynis has your number: Discover what life has in store for you through the power of numerology!* New York: Hachette Books.

Merkin, D. (2008, April 13). In search of the skeptical, hopeful, mystical Jew that could be me. *New York Times Magazine*. www.nytimes.com/2008/04/13/magazine/13kabbalah-t.html.

Muller, J. Z. (2018). *The tyranny of metrics*. Princeton, NJ: Princeton University Press.

Norman, J. M. (n.d.). The Lebombo bone, oldest known mathematical artifact. *Historyofinformation.com*. www.historyofinformation.com/detail.php?entryid=2338.

Osborn, D. (n.d.). The history of numbers. *Vedic Science*. https://vedicsciences.net/articles/history-of-numbers.html.

Pegis, R. J. (1967). Numerology and probability in Dante. *Mediaeval Studies, 29*, 370–373.

Schimmel, A. (1993). *The mystery of numbers*. New York: Oxford University Press.

Seife, C. (2010). *Proofiness: How you're being fooled by the numbers*. New York: Penguin Books.

Thimbleby, H. (2011). Interactive numbers: A grand challenge. In *Proceedings of the IADIS International Conference on Interfaces and Human Computer Interaction 2011*.

Thimbleby, H. & Cairns, P. (2017). Interactive numerals. *Royal Society Open Science*, 4(4). https://doi.org/10.1098/rsos.160903.

Wilkie, J. E. & Bodenhausen, G. (2012). Are numbers gendered? *Journal of Experimental Psychology: General, 141*(2). https://doi.org/10.1037/a0024875.

2: NUMBERS AND THE BODY

Andres, M., Davare, M., Pesenti, M., Olivier, E. & Seron, X. (2004). Number magnitude and grip aperture interaction. *Neuroreport, 15*(18), 2773–2777.

Cantlon, J. F., Brannon, E. M., Carter, E. J. & Pelphrey, K. A. (2006). Functional imaging of numerical processing in adults and 4-y-old children. *PLoS Biol, 4*(5).

Cantlon, J. F., Merritt, D. J. & Brannon, E. M. (2016). Monkeys display classic signatures of human symbolic arithmetic. *Animal Cognition, 19*(2), 405–415.

Chang, E. S., Kannoth, S., Levy, S., Wang, S. Y., Lee, J. E., et al. (2020). Global reach of ageism on older persons' health: A systematic review. *PLoS ONE, 15*(1). https://doi.org/10.1371/journal.pone.0220857.

Dehaene, S. & Changeux, J. P. (1993). Development of elementary numerical abilities: A neuronal model. *Journal of Cognitive Neuroscience, 5*(4), 390–407.

Dehaene, S., Piazza, M., Pinel, P. & Cohen, L. (2003). Three parietal circuits for number processing. *Cognitive Neuropsychology, 20*(3–6), 487–506.

DeMarree, K. G., Wheeler, S. C. & Petty, R. E. (2005). Priming a new identity: Self-monitoring moderates the effects of nonself primes on self-judgments and behavior. *Journal of Personality and Social Psychology, 89*(5), 657–671.

Fischer, M. H. (2012). A hierarchical view of grounded, embodied, and situated numerical cognition. *Cognitive Processing, 13*, 161–164.

Fischer, M. H. & Brugger, P. (2011). When digits help digits: Spatial-numerical associations point to finger counting as prime example of embodied cognition. *Frontiers in Psychology, 2*. https://doi.org/10.3389/fpsyg.2011.00260.

Gordon, P. (2004). Numerical cognition without words: Evidence from Amazonia. *Science, 306*(5695), 496–499.

Grade, S., Badets, A. & Pesenti, M. (2017). Influence of finger and mouth action observation on random number generation: An instance of embodied cognition for abstract concepts. *Psychological Research, 81*(3), 538–548.

Hauser, M. D., Tsao, F., Garcia, P. & Spelke, E. S. (2003). Evolutionary foundations of number: Spontaneous representation of numerical magnitudes by cotton-top tamarins. *Proceedings of the Royal Society of London. Series B: Biological Sciences, 270*(1523), 1441–1446.

Hubbard, E. M., Piazza, M., Pinel, P. & Dehaene, S. (2005). Interactions between number and space in parietal cortex. *Nature Reviews Neuroscience, 6*, 435–448.

Hyde, D. C. & Spelke, E. S. (2009). All numbers are not equal: An electrophysiological investigation of small and large number representations. *Journal of Cognitive Neuroscience, 21*(6), 1039–1053.

Kadosh, R. C., Lammertyn, J. & Izard, V. (2008). Are numbers special? An overview of chronometric, neuroimaging, developmental and comparative studies of magnitude representation. *Progress in Neurobiology, 84*(2), 132–147.

Lachmair, M., Ruiz Fernàndez, S., Moeller, K., Nuerk, H. C. & Kaup, B. (2018). Magnitude or multitude—what counts? *Frontiers in Psychology, 9*, 59–65.

Luebbers, P. E., Buckingham, G. & Butler, M. S. (2017). The National Football League–225 bench press test and the size-weight illusion. *Perceptual and Motor Skills, 124*(3), 634–648.

Moeller, K., Fischer, U., Link, T., Wasner, M., Huber, S., et al. (2012). Learning and development of embodied numerosity. *Cognitive Processing, 13*(1), 271–274.

Nikolova, V. (2021, August 6). Why you are 12% more likely to run a marathon at a milestone age? *Runrepeat*. https://runrepeat.com/12-percent-more-likely-to-run-a-marathon-at-a-milestone-age.

Notthoff, N., Drewelies, J., Kazanecka, P., Steinhagen-Thiessen, E., Norman, K., et al. (2018). Feeling older, walking slower—but only if someone's watching. Subjective age is associated with walking speed in the laboratory, but not in real life. *European Journal of Ageing, 15*(4), 425–433.

Pica, P., Lemer, C., Izard, V. & Dehaene, S. (2004). Exact and approximate arithmetic in an Amazonian indigene group. *Science, 306*(5695), 499–503.

Reinhard, R., Shah, K. G., Faust-Christmann, C. A. & Lachmann, T. (2020). Acting your avatar's age: Effects of virtual reality avatar embodiment on real life walking speed. *Media Psychology, 23*(2), 293–315.

Robson, D. (2018, July 19). The age you feel means more than your actual birthdate. *BBC*. www.bbc.com/future/article/20180712-the-age-you-feel-means-more-than-your-actual-birthdate.

Schwarz, W. & Keus, I. M. (2004). Moving the eyes along the mental number line: Comparing SNARC effects with saccadic and manual responses. *Perception & Psychophysics, 66*(4), 651–664.

Shaki, S. & Fischer, M. H. (2014). Random walks on the mental number line. *Experimental Brain Research, 232*(1), 43–49.

Studenski, S., Perera, S., Patel, K., Rosano, C., Faulkner, K., et al. (2011). Gait speed and survival in older adults. *Journal of the American Medical Association, 305*(1), 50–58.

Westerhof, G. J., Miche, M., Brothers, A. F., Barrett, A. E., Diehl, M., et al. (2014). The influence of subjective aging on health and longevity: A meta-analysis of longitudinal data. *Psychology and Aging, 29*(4), 793–802.

Winter, B., Matlock, T., Shaki, S. & Fischer, M. H. (2015). Mental number space in three dimensions. *Neuroscience & Biobehavioral Reviews, 57*, 209–219.

Yoo, S. C., Peña, J. F. & Drumwright, M. E. (2015). Virtual shopping and unconscious persuasion: The priming effects of avatar age and consumers' age discrimination on purchasing and prosocial behaviors. *Computers in Human Behavior, 48*, 62–71.

3: NUMBERS AND SELF-IMAGE

APS (2016, May 31). Social media "likes" impact teens' brains and behavior. *Association for Psychological Science*. www.psychologicalscience.org/news/releases/social-media-likes-impact-teens-brains-and-behavior.html.

Burrow, A. L. & Rainone, N. (2017). How many likes did I get? Purpose moderates links between positive social media feedback and self-esteem. *Journal of Experimental Social Psychology, 69*, 232–236.

Burrows, T. (2020, January 9). Social media obsessed teen who "killed herself" thought she "wasn't good enough unless she was getting likes." *The Sun*. www.thesun.co.uk/news/10705211/social-media-obsessed-death-durham-sister-tribute.

Carey-Simos, G. (2015, August 19). How much data is generated every minute on social media? *WeRSM*. https://wersm. com/how-much-data-is-generated-every-minute-on-social-media.

DNA (2020, April 20). Not able to get enough "likes" on TikTok, Noida teenager commits suicide. *DNA India*. www.dnaindia.com/india/report-not-able-to-get-enough-likes-on-tiktok-noida-teenager-commits-suicide-2821825.

Fitzgerald, M. (2019, July 18). Instagram starts test to hide number of likes posts receive for users in 7 countries. *TIME*. https://time.com/5629705/instagram-removing-likes-test.

Fliessbach, K., Weber, B., Trautner, P., Dohmen, T., Sunde, U., et al. (2007). Social comparison affects reward-related brain activity in the human ventral striatum. *Science, 318*(5894), 1305–1308.

Gaynor, G. K. (2019). Instagram removing "likes" to "depressurize" youth, some aren't buying it. *Fox News*. www.foxnews.com/lifestyle/instagram-removing-likes.

Jiang, Y., Chen, Z. & Wyer, R. S. (2014). Impact of money on emotional expression. *Journal of Experimental Social Psychology, 55*, 228–233.

Medvec V. H., Madey S. F. & Gilovich T. (October 1995). When less is more: Counterfactual thinking and satisfaction among Olympic medalists. *Journal Personality and Social Psychology, 69*(4), 603–610.

Mirror Now News (2020, April 17). Noida: Depressed over not getting enough "likes" on TikTok, youngster commits suicide. *Mirror Now Digital*. www.timesnownews.com/mirror-now/crime/article/noida-depressed-over-not-getting-enough-likes-on-tiktok-youngster-commits-suicide/579483.

Reutner, L., Hansen, J. & Greifeneder, R. (2015). The cold heart: Reminders of money cause feelings of physical coldness. *Social Psychological and Personality Science, 6*(5), 490–495.

Sherman, L. E., Payton, A. A., Hernandez, L. M., Greenfield, P. M. & Dapretto, M. (2016). The power of the Like in adolescence: Effects of peer influence on neural and behavioral responses to social media. *Psychological Science, 27*(7), 1027–1035.

Smith, K. (2019, June 1). 53 incredible Facebook statistics and facts. *Brandwatch*. www.brandwatch.com/blog/facebook-statistics.

Squires, A. (n.d.). Social media, self-esteem, and teen suicide. *PPC*. https://blog.pcc.com/social-media-self-esteem-and-teen-suicide.

Solnick, S. & Hemenway, D. (1998). Is more always better? A survey on positional concerns. *Journal of Economic Behavior & Organization, 37*(3), 373–383. https://doi.org/10.1016/S0167-2681(98)00089-4.

Vogel, E. A., Rose, J. P., Roberts, L. R. & Eckles, K. (2014). Social comparison, social media, and self-esteem. *Psychology of Popular Media Culture, 3*(4), 206–222.

Vohs, K. D. (2015). Money priming can change people's thoughts, feelings, motivations, and behaviors: An update on 10 years of experiments. *Journal of Experimental Psychology: General, 144*(4), e86–e93.

Vohs, K. D., Mead, N. L. & Goode, M. R. (2006). The psychological consequences of money. *Science, 314*(5802), 1154–1156.

Wang, S. (2019, April 30). Instagram tests removing number of "likes" on photos and videos. *Bloomberg*. www.bloomberg.com/news/articles/2019-04-30/instagram-tests-removing-number-of-likes-on-photos-and-videos.

Zaleskiewicz, T., Gasiorowska, A., Kesebir, P., Luszczynska, A. & Pyszczynski, T. (2013). Money and the fear of death: The symbolic power of money as an existential anxiety buffer. *Journal of Economic Psychology, 36*, 55–67.

4: NUMBERS AND PERFORMANCE

Ajana, B. (2018). *Metric culture: Ontologies of self-tracking practices*. Bingley, UK: Emerald Publishing.

The Economist (2019, September 12). Hugo Campos has waged a decade-long battle for access to his heart implant. Technology Quarterly. *The Economist*. www.economist.com/technology

-quarterly/2019/09/12/hugo-campos-has-waged-a-decade-long-battle-for-access-to-his-heart-implant.

Farr, C. (2015, March 17). How Tim Ferriss has turned his body into a research lab. *KQED.* www.kqed.org/futureofyou/407/how-tim-ferriss-has-turned-his-body-into-a-research-lab.

Hill, K. (2011, April 7). Adventures in self-surveillance, aka the quantified self, aka extreme navel-gazing. *Forbes.* www.forbes.com/sites/kashmirhill/2011/04/07/adventures-in-self-surveillance-aka-the-quantified-self-aka-extreme-navel-gazing/#5102dac76773.

Kuvaas, B., Buch, R. & Dysvik, A. (2020). Individual variable pay for performance, controlling effects, and intrinsic motivation. *Motivation and Emotion, 44,* 525–533.

Larsen, T. & Røyrvik, E. A. (2017). *Trangen til å telle. Objektivering, måling og standardisering som samfunnspraksis.* Oslo: Scandinavian Academic Press.

Lee, A. (2020, August 9). What is China's social credit system and why is it controversial? *South China Morning Post.* www.scmp.com/economy/china-economy/article/3096090/what-chinas-social-credit-system-and-why-it-controversial.

Lupton, D. (2016). *The quantified self.* Malden, MA: Polity Press.

Moschel, M. (2018, August 8). The beginner's guide to quantified self (plus, a list of the best personal data tools out there). *Technori.* https://technori. com/2018/08/4281-the-beginners-guide-to-quantified-self-plus-a-list-of-the-best-personal-data-tools-out-there/markmoschel.

Nafus, D. (Ed.). (2016). *Quantified: Biosensing technologies in everyday life.* Cambridge, MA: MIT Press.

Neff, G. & Nafus, D. (2016). *Self-tracking.* Cambridge, MA: MIT Press.

Quantified Self (2018, April 28). Hugo Campos: 10 years with an implantable cardiac device and "almost" no data access. *Quantified Self Public Health.* https://medium.com/quantified-self-public-health/hugo-campos-10-years-with-an-implantable-cardiac-device-and-almost-no-data-access-71018b39b938.

Ramirez, E. (2015, February 4). My device, my body, my data. *Quantified Self Public Health.* https://quantifiedself.com/blog/my-device-my-body-my-data-hugo-campos.

Satariano, A. (2020, August 4). Google faces European inquiry into Fitbit acquisition. *New York Times*. www.nytimes.com/2020/08/04/business/google-fitbit-europe.html.

Selke, S. (Ed.). (2016). *Lifelogging: Digital self-tracking and lifelogging—between disruptive technology and cultural transformation*. Wiesbaden: Springer VS.

Stanford Medicine X (n.d.). Hugo Campos. *Stanford Medicine X*. https://medicinex.stanford.edu/citizen-campos.

5: NUMBERS AND EXPERIENCES

Dijkers, M. (2010). Comparing quantification of pain severity by verbal rating and numeric rating scales. *Journal of Spinal Cord Medicine, 33*(3), 232–242.

Erskine, R. (2018, May 15). You just got attacked by fake 1-star reviews. Now what? *Forbes*. www.forbes.com/sites/ryanerskine/2018/05/15/you-just-got-attacked-by-fake-1-star-reviews-now-what/#5c0b23cc1071.

Hoch, S. J. (2002). Product experience is seductive. *Journal of Consumer Research, 29*(3), 448–454.

Liptak, A. (2018, February 2). Facebook strikes back against the group sabotaging Black Panther's Rotten Tomatoes rating. *The Verge*. www.theverge.com/2018/2/2/16964312/facebook-black-panther-rotten-tomatoes-last-jedi-review-bomb.

Rockledge, M. D., Rucker, D. D. & Nordgren, L. F. (2021, April 8). Mass-scale emotionality reveals human behaviour and marketplace success. *Nature Human Behaviour, 5*, 1323–1329.

Williamson, A. & Hoggart, B. (2005). Pain: A review of three commonly used pain rating scales. *Journal of Clinical Nursing, 14*(7), 798–804.

6: NUMBERS AND RELATIONSHIPS

American Psychological Association (APA) (2016, August 4). Tinder: Swiping self esteem? *APA*. www.apa.org/news/press/releases/2016/08/tinder-self-esteem.

Danaher, J., Nyholm, S. & Earp, B. D. (2018). The quantified relationship. *American Journal of Bioethics, 18*(2), 3–19.

Eurostat (2018, July 6). Rising proportion of single person households in the EU. *Eurostat.* https://ec.europa.eu/eurostat/web/products-eurostat-news/-/ddn-20180706-1.

Ortiz-Ospina, E. & Roser, M. (2016). Trust. *Our World in Data.* https://ourworldindata.org/trust.

Strubel, J. & Petrie, T. A. (2017). Love me Tinder: Body image and psychosocial functioning among men and women. *Body Image, 21,* 34–38.

Timmermans, E., De Caluwé, E. & Alexopoulos, C. (2018). Why are you cheating on Tinder? Exploring users' motives and (dark) personality traits. *Computers in Human Behavior, 89,* 129–139.

Waldinger, M. D., Quinn, P., Dilleen, M., Mundayat, R., Schweitzer, D. H., et al. (2005). A mutinational population survey of intravaginal ejaculation latency time. *Journal of Sexual Medicine, 2*(4), 492–497.

Ward, J. (2017). What are you doing on Tinder? Impression management on a matchmaking mobile app. *Information, Communication & Society, 20*(11), 1644–1659.

Wellings, K., Palmer, M. J., Machiyama, K. & Slaymaker, E. (2019). Changes in, and factors associated with, frequency of sex in Britain: Evidence from three national surveys of sexual attitudes and lifestyles (Natsal). *British Medical Journal, 365*(8198).

World Values Survey (WVS) (n.d.). Online data analysis. *WVS.* www.worldvaluessurvey.org/WVSOnline.jsp.

7: NUMBERS AS CURRENCY

Barlyn, S. (2018, September 19). Strap on the Fitbit: John Hancock to sell only interactive life insurance. *Reuters.* www.reuters.com/article/us-manulife-financi-john-hancock-lifeins-idUSKCN1LZ1WL.

BBC News (2018, September 20). John Hancock adds fitness tracking to all policies. *BBC News.* www.bbc.com/news/technology-45590293.

Blauw, S. (2020). *The number bias: How numbers lead and mislead us.* London: Hodder & Stoughton.

Brown, A. (2020, August 6). TikTok's 7 highest-earning stars: New Forbes list led by teen queens Addison Rae and Charli D'Amelio. *Forbes*. www.forbes.com/sites/abrambrown/2020/08/06/tiktoks-highest-earning-stars-teen-queens-addison-rae-and-charli-damelio-rule/?sh=2e41abf75087.

The Ezra Klein Show (2022, February 25). Transcript: Ezra Klein interviews C. Thi Nguyen. *New York Times*. www.nytimes.com/2022/02/25/podcasts/transcript-ezra-klein-interviews-c-thi-nguyen.html.

Frazier, L. (2020, August 10). 5 ways people can make serious money on TikTok. *Forbes*. www.forbes.com/sites/lizfrazierpeck/2020/08/10/5-ways-people-can-make-serious-money-on-tiktok/?sh=19aea32a5afc.

Meyer, R. (2015, September 25). Could a bank deny your loan based on your Facebook friends? *The Atlantic*. www.theatlantic.com/technology/archive/2015/09/facebooks-new-patent-and-digital-redlining/407287.

Nguyen, C. Thi (2020). *Games: Agency as Art*. New York: Oxford University Press.

Nødtvedt, K. B., Sjåstad, H., Skard, S. R., Thorbjørnsen, H. & Van Bavel, J. J. (2021, April 29). Racial bias in the sharing economy and the role of trust and self-congruence. *Journal of Experimental Psychology: Applied*, *27*(3), 508–528.

Wang, L., Zhong, C. B. & Murnighan, J. K. (2014). The social and ethical consequences of a calculative mindset. *Organizational Behavior and Human Decision Processes*, *125*(1), 39–49.

8: NUMBERS AND THE TRUTH

Bhatia, S., Walasek, L., Slovic, P. & Kunreuther, H. (2021). The more who die, the less we care: Evidence from natural language analysis of online news articles and social media posts. *Risk Analysis*, *41*(1), 179–203.

Henke, J., Leissner, L. & Möhring, W. (2020). How can journalists promote news credibility? Effects of evidences on trust and credibility. *Journalism Practice*, *14*(3), 299–318.

Koetsenruijter, A. W. M. (2011). Using numbers in news increases story credibility. *Newspaper Research Journal, 32*(2), 74–82.

Lindsey, L. L. M. & Yun, K. A. (2003). Examining the persuasive effect of statistical messages: A test of mediating relationships. *Communication Studies, 54*(3), 306–321.

Luo, M., Hancock, J. T. & Markowitz, D. M. (2020). Credibility perceptions and detection accuracy of fake news headlines on social media: Effects of truth-bias and endorsement cues. *Communication Research, 49*(2), 171–195.

Luppe, M. R. & Lopes Fávero, L. P. (2012). Anchoring heuristic and the estimation of accounting and financial indicators. *International Journal of Finance and Accounting, 1*(5), 120–130.

Peter, L. (2022, May 1). How Ukraine's "Ghost of Kyiv" Legendary Pilot was born, BBC News. https://www.bbc.com/news/world-europe-61285833.

Plous, S. (1989). Thinking the unthinkable: The effects of anchoring on likelihood estimates of nuclear war. *Journal of Applied Social Psychology, 19*(1), 67–91.

Seife, C. (2010). *Proofiness: How you're being fooled by the numbers*. New York: Penguin books.

Slovic, S. & Slovic, P. (2015). *Numbers and nerves: Information, emotion, and meaning in a world of data*. Corvallis: Oregon State University Press.

Tomm, B. M., Slovic, P. & Zhao, J. (2019). The number of visible victims shapes visual attention and compassion. *Journal of Vision, 19*(10), 105.

Van Brugen, I. (2022, February 25). Who is the Ghost of Kyiv? Ukraine MiG-29 Fighter Pilot Becomes the Stuff of Legend, *Newsweek*. https://www.newsweek.com/who-ghost-kyiv-ukraine-fighter-pilot-mig-29-russian-fighter-jets-combat-1682651.

Yamagishi, K. (1997). Upward versus downward anchoring in frequency judgments of social facts. *Japanese Psychological Research, 39*(2), 124–129.

Ye, Z., Heldmann, M., Slovic, P. & Münte, T. F. (2020). Brain imaging evidence for why we are numbed by numbers. *Scientific Reports, 10*(1). www.nature.com/articles/s41598-020-66234-z.

9: NUMBERS AND SOCIETY

Alexander, M. & Fisher, T. (2003). Truth and consequences: Using the bogus pipeline to examine sex differences in self-reported sexuality. *Journal of Sex Research*, *40*(1), 27–35. https://doi.org/10.1080/00224490309552164.

Ariely, D., Loewenstein G. & Prelec, D. (2003). "Coherent arbitrariness": Stable demand curves without stable preferences. *Quarterly Journal of Economics*, *118*(1), 73–105.

Bevan, G. & Hood, C. (2006): What's measured is what matters: Targets and gaming in the English public health system. *Public Administration* 84(3).

Blauw, S. (2020). *The number bias: How numbers lead and mislead us*. London: Hodder & Stoughton.

Brennan, L., Watson, M., Klaber, R. & Charles, T. (2012). The importance of knowing context of hospital episode statistics when reconfiguring the NHS. *British Medical Journal*, *344*, e2432. https://doi.org/10.1136/bmj.e2432.

Campbell, S. D. & Sharpe, S. A. (2009). Anchoring bias in consensus forecasts and its effect on market prices. *Journal of Financial and Quantitative Analysis*, *44*(2), 369–390.

Chan, A. (2013, May 30). 1998 study linking autism to vaccines was an "elaborate fraud." *Live Science*. www.livescience.com/35341-mmr-vaccine-linked-autism-study-was-elaborate-fraud.html.

Chatterjee, P. & Joynt, K. E. (2014). Do cardiology quality measures actually improve patient outcomes? *Journal of the American Heart Association*, February. https://doi.org/10.1161/JAHA.113.000404.

Dunn, T. (2016, August 10). 11 ridiculous future predictions from the 1900 World's Fair—and 3 that came true. *Upworthy*. www.upworthy.com/11-ridiculous-future-predictions-from-the-1900-worlds-fair-and-3-that-came-true.

Financial Times (2016, April 14). How politicians poisoned statistics. *Financial Times*. www.ft.com/content/2e43b3e8-01c7-11e6-ac98-3c15a1aa2e62.

Fliessbach, K., Weber, B., Trautner, P., Dohmen, T., Sunde, U., et al.

(2007). Social comparison affects reward-related brain activity in the human ventral striatum. *Science, 318*(5854), 1305–1308.

Furnham, A. & Boo, H. C. (2011). A literature review of the anchoring effect. *Journal of Socio-economics, 40*(1), 35–42.

Gwiazda, J., Ong, E., Held, R. & Thorn F. (2000). Myopia and ambient night-time lighting. *Nature, 404*, 144.

Hans, V. P., Helm, R. K. & Reyna, V. F. (2018). From meaning to money: Translating injury into dollars. *Law and Human Behavior, 42*(2), 95–109.

Hviid, A., Hansen, J. V., Frisch, M. & Melbye, M. (2019). Measles, mumps, rubella vaccination and autism: A nationwide cohort study. *Annals of Internal Medicine, 170*(8), 513–520.

Johnson, E. & Goldstein, D. (2003). Do defaults save lives? *Science, 302*(5649), 1338–1339. https://doi.org/10.1126/science.1091721.

Kahan, D. M., Peters, E., Cantrell Dawson, E. & Slovic, P. (2017). Motivated numeracy and enlightened self-government. *Behavioural Public Policy, 1*(1), 54–86.

King, A. (2016, November 12). Poll expert eats bug after being wrong about Trump. *CNN Politics*. https://edition.cnn.com/2016/11/12/politics/pollster-eats-bug-after-donald-trump-win/index.html.

Lalot, F., Quiamzade, A. & Falomir-Pichastor, J. M. (2019). How many migrants are people willing to welcome into their country? The effect of numerical anchoring on migrant acceptance. *Journal of Applied Social Psychology, 49*(6), 361–371.

Larsen, T. & Røyrvik, E. A. (2017). *Trangen til å telle. Objektivering, måling og standardisering som samfunnspraksis*. Oslo: Scandinavian Academic Press.

Lee, S. (2018, February 25). Here's how Cornell scientist Brian Wansink turned shoddy data into viral studies about how we eat. *Buzzfeed News*. www.buzzfeednews.com/article/stephaniemlee/brian-wansink-cornell-p-hacking.

Mau, S. (2019). *The metric society: On the quantification of the social*. Medford, MA: Polity Press.

Muller, J. Z. (2018). *The tyranny of metrics*. Princeton, NJ: Princeton University Press.

National Geographic (2011). Y2K bug. *National Geographic*. www.nationalgeographic.org/encyclopedia/Y2K-bug.

OECD (2020). OECD employment outlook 2020: Worker security and the COVID-19 crisis. *OECD*. www.oecd.org/employment-outlook/2020.

Ohio State University (1999). Night lights don't lead to nearsightedness, study suggests. *ScienceDaily*. www.sciencedaily.com/releases/2000/03/000309074442.htm.

Quinn, G., Shin, C., Maguire, M., et al. (1999). Myopia and ambient lighting at night. *Nature, 399*(6732), 113–114. https://doi.org/10.1038/20094.

Schofield, J. (2000, January 5). The Millennium bug: Special report. *The Guardian*. www.theguardian.com/technology/2000/jan/05/y2k.guardiananalysispage.

Seife, C. (2010). *Proofiness: How you're being fooled by the numbers*. New York: Penguin books.

Spiegelhalter, D. (2015). *Sex by numbers*. London: Profile Books.

Tamma, P. D., Ault, K. A., del Rio, C., Steinhoff, M. C., Halsey, N. A., et al. (2009). Safety of influenza vaccination during pregnancy. *American Journal of Obstetrics and Gynecology, 201*(6), 547–552. https://doi.org/10.1016/j.ajog.2009.09.034.

Tversky, A. & Kahneman, D. (1974). Judgment under uncertainty: Heuristics and biases. *Science, 185*(4157), 1124–1131.

Vogel, E. A., Rose, J. P., Roberts, L. R. & Eckles, K. (2014). Social comparison, social media, and self-esteem. *Psychology of Popular Media Culture, 3*(4), 206–222.

Zadnik, K., Jones, L., Irvin, B., Kleinstein, R., Manny, R., et al. (2000). Myopia and ambient night-time lighting. *Nature, 404*, 143–144.

10: NUMBERS AND YOU

Brendl, C. M., Markman, A. B. & Messner, C. (2003). The devaluation effect: Activating a need devalues unrelated objects. *Journal of Consumer Research, 29*(4), 463–473.

Castro, J. (2014, January 30). When was Jesus born? *Live Science*. www.livescience.com/42976-when-was-jesus-born.html.

Dohmen, T. J., Falk, A., Huffman, D. & Sunde, U. (2006). Seemingly irrelevant events affect economic perceptions and expectations: The FIFA World Cup 2006 as a natural experiment. *IZA Institute of Labor Economics*. www.iza.org/publications/dp/2275/seemingly-irrelevant-events-affect-economic-perceptions-and-expectations-the-fifa-world-cup-2006-as-a-natural-experiment.

Friberg, R. & Mathä, T. Y. (2004). Does a common currency lead to (more) price equalization? The role of psychological pricing points. *Economics Letters, 84*(2), 281–287.

Kämpfer, S. & Mutz, M. (2013). On the sunny side of life: Sunshine effects on life satisfaction. *Social Indicators Research, 110*(2), 579–595.

Knapton, S. (2020, October 6). An earlier universe existed before the big bang, and can still be observed today, says Nobel winner. *The Telegraph*. www.telegraph.co.uk/news/2020/10/06/earlier-universe-existed-big-bang-can-observed-today.

Kumar, M. (2019, May 15). When maths goes wrong. *New Statesman*. www.newstatesman.com/culture/books/2019/05/when-maths-goes-wrong.

Raghubir, P. & Srivastava, J. (2002). Effect of face value on product valuation in foreign currencies. *Journal of Consumer Research, 29*(3), 335–347.

Schwarz, N., Strack, F., Kommer, D. & Wagner, D. (1987). Soccer, rooms, and the quality of your life: Mood effects on judgments of satisfaction with life in general and with specific domains. *European Journal of Social Psychology, 17*(1), 69–79.

Tom, G. & Rucker, M. (1975). Fat, full, and happy: Effects of food deprivation, external cues, and obesity on preference ratings, consumption, and buying intentions. *Journal of Personality and Social Psychology, 32*(5), 761–766.

INDEX

Index